Learning from Accidents

Learning from Accidents

Second edition

Trevor Kletz
DSc, FEng, FRSC, FIChemE

Butterworth-Heinemann Ltd
Linacre House, Jordan Hill, Oxford OX2 8DP

 A member of the Reed Elsevier group

OXFORD   LONDON   BOSTON
MUNICH   NEW DELHI   SINGAPORE   SYDNEY
TOKYO   TORONTO   WELLINGTON

First published as *Learning from Accidents in Industry* 1988
Reprinted 1990
Second edition 1994

© Butterworth-Heinemann Ltd 1994

British Library Cataloguing in Publication Data
Kletz, Trevor A.
  Learning from Accidents. – 2Rev.ed
  I. Title
  363.117072

ISBN 0 7506 1952 X

Library of Congress Cataloging in Publication Data
Kletz, Trevor A.
  Learning from accidents/Trevor Kletz. – 2nd ed.
  p.   cm.
  Includes bibliographical references and index.
  ISBN 0 7506 1952 X
  1. Chemical industry – Accidents. 2. Industrial accidents.
  3. Industrial accidents – Investigation. I. Title.
  HD7269.C45K43   93–50876
  363.1'165–dc20   CIP

Composition by Scribe Design, Gillingham, Kent
Printed and bound in Great Britain by Redwood Books, Trowbridge

Contents

Forethoughts

It is the success of engineering which holds back the growth of engineering knowledge, and its failures which provide the seeds for its future development.

D. I. Blockley and J. R. Henderson, *Proc. Inst. Civ. Eng.* Part 1, Vol 68, Nov. 1980, p. 719.

What has happened before will happen again. What has been done before will be done again. There is nothing new in the whole world.

Ecclesiastes, 1, 9 (*Good News Bible*).

What worries me is that I may not have seen the past here – perhaps I have seen the future.

Elie Wiesel

Below, distant, the roaring courtiers
rise to their feet – less shocked than irate.
Salome has dropped the seventh veil
and they've discovered there are eight.

Danny Abse, *Way out in the Centre.*

... But if so great desire
Moves you to hear the tale of our disasters
Briefly recalled ...
However I may shudder at the memory
And shrink again in grief, let me begin.

Virgil, *The Aeneid.*

Preface

I would like to thank the companies where the accidents I have described occurred for letting me publicise their failures, so that others can learn from them, and the many colleagues with whom I have discussed these accidents and who made various comments, some emphasising the immediate causes and others the underlying ones. All were valuable. My colleagues – particularly those who attended the discussions described in Part 4 of the Introduction – are the real authors of this book. I am merely the amanuensis.

Rabbi Judah the Prince (c. 135–217 AD) said, 'Much have I learnt from my teachers, more from my colleagues and most of all from my students'.

I do not always name the products made on the plants where the incidents occurred, partly to preserve their anonymity but also for another reason: If I said that an explosion occurred on a plant manufacturing acetone, readers who do not use acetone might be tempted to ignore that report. In fact, most of the recommendations apply to most plants, regardless of the materials they handle. To misquote the well-known words of the poet John Donne,

> No plant is an Island, entire of itself; every plant is a piece of the Continent,
> a part of the main. Any plant's loss diminishes us, because we are involved
> in the Industry; and therefore never send to know for whom the inquiry
> sitteth; it sitteth for thee.

Descriptions of most of the accidents described in this book have appeared before but scattered throughout various publications, often in a different form. References are given at the end of each chapter and thanks are due to the original publishers for permission to quote from them.

For this second edition I have added chapters on some of the major incidents that have occurred since the first edition was written and I have made some changes and additions to the original text. I have retained the original chapter numbers, except that the last chapter is now number twenty-two. I am grateful to Brian Appleton, one of the assessors at the Piper Alpha inquiry, for writing a chapter on that accident.

Since the first edition was published I have written a book with a rather similar title, *Lessons from Disaster – How Organisations have No Memory and Accidents Recur* (Institution of Chemical Engineers, 1993) but its theme is rather different. This book deals mainly with accident investigation and the need to look beyond the immediate technical causes for ways of avoiding the hazards and for weaknesses in the management system. The other book, as the sub-title indicates, shows how accidents are forgotten and then repeated, and suggests ways of improving the corporate memory.

To avoid the clumsy phrases 'he or she' and 'his or hers' I have usually used 'he' or 'his'. There has been a welcome increase in the number of women working in industry but the manager, designer or accident victim is still usually male.

A note for American readers

The term 'plant manager' is used in the UK sense to describe the first level of professional management, someone who would be known as a supervisor in most US companies. The person in charge of a site is called a works manager.

A note on units

I have used the units likely to be most familiar to the majority of my readers.

Short lengths, such as pipeline sizes, are in inches and millimetres; longer lengths are in metres only (1 metre = 3.28 feet).

Pressures are in pounds force per square inch (psi) and bars (1 bar = 100 kilopascals and is also almost exactly 1 atmosphere and 1 kilogram per square centimetre).

Masses are in kilograms or metric tonnes (1 metric tonne = 1.10 short [US] tons or 0.98 long [UK] tons).

Volumes are in cubic metres (1 m^3 = 264 US gallons or 220 imperial gallons or 35.3 cubic feet).

Temperatures are in degrees Celsius (°C).

A note on the organisation of maintenance in the process industries

A note on this subject may be helpful to readers from other industries. In most process industry factories, including oil refineries and chemical works, there is a dual organisation. One stream of managers, foremen and operators are responsible for running the process while another stream of engineers, foremen and craftsmen are responsible for repairs. The two streams meet in the person of the factory manager. When repairs or overhauls are necessary the process team prepare the equipment, usually by isolating it and removing hazardous materials, and then hand it over to the maintenance team. This is usually done by completion of a permit-to-work which describes the work to be done, any remaining hazards and the precautions necessary. It is prepared by a process foreman or senior operator and accepted by the craftsman who is going to carry out the maintenance or his foreman. When the repairs are complete the permit is returned and then, but not before, the plant can be started up.

Many accidents have occurred because the permit system was poor or was not followed correctly (see Chapters 2, 5 and 17).

At times companies have experimented with 'manageers', people who combined the jobs of manager (of the process) and maintenance engineer. On the whole such marriages have not been a success, as few people have the knowledge and experience needed to carry out two such different tasks.

Acknowledgements

Thanks are due to the companies where the accidents, described in this book, occurred for permission to describe them, so that we may all learn from them, to the Leverhulme Trust for financial support, to Loughborough University of Technology for giving me the opportunity to develop and record some of the knowledge I acquired during my thirty-eight years in the chemical industry, to Professor F. P. Lees who read the book in manuscript and made many valuable suggestions, and to Mr E. S. Hunt for assistance with Chapter 15.

The book is dedicated to all those killed or injured in the accidents, in the hope that others will learn from their misfortunes.

Introduction

Find a little, learn a lot. An archaeological magazine[1]

Accident investigation is like peeling an onion or, if you prefer more poetic metaphors, dismantling a Russian doll or the dance of the seven veils. Beneath one layer of causes and recommendations there are other, less superficial layers. The outer layers deal with the immediate technical causes while the inner layers are concerned with ways of avoiding the hazards and with the underlying causes, such as weaknesses in the management system. Very often only the outer layers are considered and thus we fail to use all the information for which we have paid the high price of an accident. The aim of this book is to show, by analysing accidents that have occurred, how we can learn more from accidents and thus be better able to prevent them occurring again. The incidents discussed range from the trivial to major accidents such as Flixborough, Bhopal and Piper Alpha. The first edition discussed accidents which had occurred mainly in the chemical industry, but this second edition covers a wider range. The book should therefore interest all those concerned with the investigation of accidents, of whatever sort, and all those who work in industry, whether in design, operations or loss prevention.

I am not suggesting that the immediate causes of an accident are any less important than the underlying causes. All must be considered if we wish to prevent further accidents, as the examples will show. But putting the immediate causes right will prevent only the last accident happening again; attending to the underlying causes may prevent many similar accidents.

Compared with some other books on accidents (for example, reference 2) I have emphasised cause and prevention rather than human interest or cleaning up the mess. I have taken it for granted that my readers are fully aware of the suffering undergone by the bereaved and injured and that there is no need for me to spell it out. If we have not always prevented accidents in the past this is due to lack of knowledge, not lack of desire.

1   Finding the facts

This book is not primarily concerned with the collection of information about accidents but with the further consideration of facts already collected. Those interested in the collection of information should consult a book by the Center for Chemical Process Safety[3], a paper by Craven[4] or, if sabotage is suspected, papers by Carson and Mumford[5].

Nevertheless, it may be useful to summarise a few points that are sometimes overlooked[6].

(1) The investigating panel should not be too large, four or five people are usually sufficient, but should include people with a variety of experience and at least one person from another part of the organisation. Such a person is much more likely than those closely involved to see the wider issues and the relevance of the incident to other plants. It is difficult to see the shape of the forest when we are in the middle of it.

(2) Try not to disturb evidence that may be useful to experts who may be called in later. If equipment has to be moved, for example, to make the plant safe, then photograph it first. In the UK a factory inspector may direct that things are left undisturbed 'for so long as is reasonably necessary for the purpose of any examination or investigation'.

(3) Draw up a list of everyone who may be able to help, such as witnesses, workers on other shifts, designers, technical experts, etc. Interview witnesses as soon as you can, before their memories fade and the story becomes simpler and more coherent.

(4) Be patient when questioning witnesses. Let people ramble on in a relaxed manner. Valuable information may be missed if we try to take police-type statements.

Do not question witnesses in such a way that you put ideas into their minds. Try to avoid questions to which the answer is 'yes' or 'no'. It is easier for witnesses to say 'yes' or 'no' than to enter into prolonged discussions, especially if they are suffering from shock.

(5) Avoid, at this stage (preferably at any stage; see later), any suggestion of blame. Make it clear that the objective of the investigation is to find out the facts, so that we can prevent the accident happening again. An indulgent attitude towards people who have had lapses of attention, made errors of judgement or not always followed the rules is a price worth paying in order to find out what happened.

(6) Inform any authorities who have to be notified (in the UK a wide variety of dangerous occurrences have to be notified to the Health and Safety Executive under *The Reporting of Injuries, Diseases and Dangerous Occurrences Regulations 1985*) and the insurance company, if claims are expected.

(7) Record information, quantitative if possible, on damage and injuries so that others can use it for prediction.

Ferry[7] and Lynch[8] give more guidance on the collection of the facts.

2   Avoid the word 'cause'

Although I have used this word it is one I shall use sparingly when analysing accidents, for four reasons.

(1) If we talk about causes we may be tempted to list those we can do little or nothing about. For example, a source of ignition is often said to be the cause of a fire. But when flammable vapour and air are mixed in the flammable range, experience shows that a source of ignition is liable to turn up, even though we have done everything possible to remove known sources of ignition (see Chapter 4.) The only really effective way of preventing an ignition is to prevent leaks of flammable vapour. Instead of asking, 'What is the cause of this fire?' we should ask 'What is the most effective way of preventing another similar fire?' We may then think of ways of preventing leaks.

Another example: Human error is often quoted as the cause of an accident but as I try to show in my book, *An Engineer's View of Human Error*[9], there is little we can do to prevent people making errors, especially those due to a moment's forgetfulness. If we ask 'What is the cause of this accident?' we may be tempted to say 'Human error' but if we ask 'What should we do differently to prevent another accident?' we are led to think of changes in design or methods of operation (see Section 22.8).

(2) The word 'cause' has an air of finality about it that discourages further investigation. If a pipe fails, for example, and the cause is said to be corrosion we are tempted to think that we know why it failed. But to say that a pipe failure was due to corrosion is rather like saying that a fall was due to gravity. It may be true but it does not help us to prevent further failures. We need to know the answers to many more questions: Was the material of construction specified correctly? Was the specified material actually used? Were operating conditions the same as those assumed by the designers? What corrosion monitoring did they ask for? Was it carried out? Were the results ignored? And so on.

(3) The word 'cause' implies blame and people become defensive. So instead of saying that an accident was caused by poor design (or maintenance or operating methods) let us say that it could be prevented by better design (or maintenance or operating methods). We are reluctant to admit that we did something badly but we are usually willing to admit that we could do it better.

(4) If asked for the cause of an accident people often suggest abstractions such as institutional failure, new technology, Acts of God or fate. But institutions and technology have no minds of their own and cannot

change on their own: someone has to do something. We should say who and what. Lightning and other so-called Acts of God cannot be avoided but we know they will occur and blaming them is about as helpful as blaming daylight or darkness. Fate is just a lazy person's excuse for doing nothing.

However, the main point I wish to make is that whether we talk about causes or methods of prevention, we should look below the immediate technical changes needed, at the more fundamental changes such as ways of avoiding the hazard and ways of improving their management system.

3   The irrelevance of blame

If accident investigations are conducted with the objective of finding culprits and punishing them, then people do not report all the facts, and who can blame them? We never find out what really happened and are unable to prevent it happening again. If we want to know what happened we have to make it clear that the objective of the inquiry is to establish the facts and make recommendations and that nobody will be punished for errors of judgement or for forgetfulness, only for deliberate, reckless or repeated indifference to the safety of others. Occasional negligence may go unpunished, but this is a small price to pay to prevent further accidents. An accident may show that someone does not have the ability to carry out a particular job and he may have to be moved, but this is not punishment and should not be made to look like punishment.

In fact very few accidents are the result of negligence. Most human errors are the result of a moment's forgetfulness or aberration, the sort of error we all make from time to time. Others are the result of errors of judgement, inadequate training or instruction or inadequate supervision[9].

Accidents are rarely the fault of a single person. Responsibility is usually spread amongst many people. To quote from an official UK report on safety legislation[10]:

> The fact is – and we believe this to be widely recognised – the traditional concepts of the criminal law are not readily applicable to the majority of infringements which arise under this type of legislation. Relatively few offences are clear cut, few arise from reckless indifference to the possibility of causing injury, few can be laid without qualification at the door of a single individual. The typical infringement or combination of infringements arises rather through carelessness, oversight, lack of knowledge or means, inadequate supervision, or sheer inefficiency. In such circumstances the process of prosecution and punishment by the criminal courts is largely an irrelevancy. The real need is for a constructive means of ensuring that practical improvements are made and preventative measures adopted.

In addition, as we shall see, a dozen or more people have opportunities to prevent a typical accident and it is unjust to pick on one of them, often the last and most junior person in the chain, and make him the scapegoat.

The views I have described are broadly in agreement with those of the UK Health and Safety Executive. They prosecute, they say, only 'when employers and others concerned appear deliberately to have disregarded the relevant regulations or where they have been reckless in exposing people to hazard or where there is a record of repeated infringement[11].' They usually prosecute the company rather than an individual because responsibility is shared by so many individuals.

4   How can we encourage people to look for underlying causes?

First they must be convinced that the underlying causes are there and that it will be helpful to uncover them. Reading this book may help. A better way is by discussion of accidents that have occurred and the action needed to prevent them happening again. The discussion leader describes an accident very briefly; those present question him to establish the rest of the facts and then say what *they think* ought to be done to prevent it happening again. The UK Institution of Chemical Engineers provides sets of notes and slides for use in such discussions[12]. The incidents in this book may also be used. It is better, however, to use incidents which have occurred in the plant in which those present normally work. Some discussion groups concentrate on the immediate causes of the incidents discussed; the discussion leader should encourage them to look also at the wider issues.

After a time, it becomes second nature for people who have looked for the less obvious ways of preventing accidents, either in discussion or in real situations, to continue to do so, without prompting.

Most of the recommendations described in this book were made during the original investigation but others only came to light when the accidents were later selected for discussion in the way I have described.

In the book the presentations differ a little from chapter to chapter, to avoid monotony and to suit the varying complexity of the accounts. Thus in discussing fires and explosions, a discussion of the source of ignition may be followed by recommendations for eliminating it. In other cases, all the facts are described first and are followed by all the recommendations.

Occasionally questions are asked to which there are no clear or obvious answers.

5   Is it helpful to use an accident model?

Many people believe that it is and a number of models have been described. For example, according to Houston[13,14] three input factors are necessary for an accident to occur: target, driving force and trigger. For example, consider a vessel damaged by pressurisation with compressed air at a pressure above the design pressure (as in the incident described in Chapter 7). The driving force is compressed air, the target is the vessel to which it is connected and the trigger is the opening of the connecting valve. The development of the accident is determined by a number of parameters: the contact probability (the probability that all the necessary input factors are present), the contact efficiency (the fraction of the driving force which reaches the target) and the contact time. The model indicates a number of ways in which the probability or severity of the accident may be reduced. One of the input factors may be removed or the effects of the parameters minimised. Pope[15] and Ramsey[16] have described other models.

Personally I have not found such models useful. I find that time may be spent struggling to fit the data into the framework and that this distracts from the free-ranging thinking required to uncover the less obvious ways of preventing the accident. A brainstorming approach is needed. I do give in the Appendix to Chapter 22 a list of questions that may help some people to look below the surface but they are in no sense a model. Use models by all means if you find them useful but do not become a slave to them. Disregard them if you find that they are not helping you.

However, although I do not find a general model useful, I do find it helpful to list the chain of events leading up to an accident and these chains are shown for each accident that is discussed in detail. They show clearly that the chain could have been broken, and the accident prevented, at any point. At one link in the chain the senior managers of the company might have prevented the accident by changing their organisation or philosophy; at another link the operator or craftsman might have prevented it by last-minute action; designers, managers and foremen also had their opportunities. The chains remind us that we should not use inaction by those above (or below) us as an excuse for inaction on our part. The explosion described in Chapter 4 would not have occurred if the senior managers had been less insular. Equally it would not have occurred if a craftsman had made a joint with greater skill.

The chain diagrams use different typefaces to illustrate the onion effect. Attention to the underlying causes may break the chain at various points, not just at the beginning, as the diagrams will show.

6   There are no right answers

If the incidents described in this book are used as subjects for discussion, as described earlier, it must be emphasised that there are no right answers for the group to arrive at. The group may think that my recommendations go too far, or not far enough, and they may be right. How far we should go is a matter of opinion. What is the right action in one company may not be right for another which has a different culture or different working practices. I have not tried to put across a set of answers for specific problems, a code or a standard method for investigating accidents but rather a way of looking at them. I have tried to preserve the divergence of view which is typical of the discussions at many inquiries so that the book has something of an oral character.

While the primary purpose of the book is to encourage people to investigate accidents more deeply, I hope that the specific technical information given in the various chapters will also be useful, in helping readers deal with similar problems on their own plants. You may not agree with my recommendations; if so, I hope you will make your own. Please do not ignore the problems. The incidents discussed did not have exotic causes, few have, and similar problems could arise on many plants. After most of them people said, 'We ought to have thought of that before'.

7 Prevention should come first

The investigations described in this book should ideally have been carried out when the plants were being designed so that modifications, to plant design or working methods, could have been made before the accidents occurred, rather than after. Samuel Coleridge described history as a lantern on the stern, illuminating the hazards the ship has passed through rather than those that lie ahead. It is better to see the hazards afterwards than not see them at all, as we may pass the same way again, but it is better still to see them when they still lie ahead. There are methods available which can help us to foresee hazards but they are beyond the scope of this book. Briefly, those that I consider most valuable are:

- Hazard and operability studies (hazops)[17,18,19] at the detailed design stage.
- A variation of the technique at the earlier, conceptual design stage[20] when we decide which product to make and by which route (see Chapter 22).
- Detailed inspection during and after construction to make sure that the design has been followed and that details not specified in the design have been constructed in accordance with good engineering practice (see Chapter 16).
- Safety audits on the operating plant[21,22].

8 Record all the facts

Investigating teams should place on record all the information they collect and not just that which they use in making their recommendations.

Readers with a different background, experience or interests may then be able to draw additional conclusions from the evidence (as shown in Chapter 14). As already stated, outsiders may see underlying causes more clearly than those who are involved in the detail. UK official reports are usually outstanding in this respect. The evidence collected is clearly displayed, then conclusions are drawn and recommendations made. Readers may draw their own conclusions, if they wish to do so. In practice they rarely draw contradictory conclusions but they may draw additional and deeper ones.

The historian, Barbara Tuchman, has written, 'Leaving things out because they do not fit is writing fiction, not history'[23].

It is usual in scholarly publications to draw all the conclusions possible from the facts. Compare, for example, the way archaeologists draw pages of deductions from a few bits of pottery ('we find one jar handle with three inscribed letters, and already "It's a literate society"'[24]). In this respect most writing on accidents has not been scholarly, authors often being content to draw only the most obvious messages.

Nevertheless reports should not be too verbose or busy people will not read them. (Chapter 15, Appendix reproduces a good report.) The ideal is two reports: one giving the full story and the other summarising the events and drawing attention to those recommendations of general interest which apply outside the unit where the incident occurred.

9   Other information to include in accident reports

We should include the following information in accident reports, but often do not:

- **Who** is responsible for carrying out the recommendations? Nothing will be done unless someone is clearly made responsible.

  Each works or department should have a procedure for making sure that they consider recommendations from other works and departments. In particular, design departments should have a procedure for making sure that they consider recommendations made by the works. Sometimes these are ignored because they are impracticable or because the designers resent other people telling them how to do their job. Any recommendations for changes in design codes or procedures should be discussed with the design department before issue.
- **When** will the recommendations be complete? The report can then be brought forward at this time.
- **How much** will they cost, in money and other resources (for example, two design engineers for three weeks or one electrician for three days)? We can then see if the resources are likely to be available. In addition, though safety is important, we should not write blank cheques after an accident. If the changes proposed are expensive we should ask if the

risk justifies the expenditure or if there is a cheaper way of preventing a recurrence. The law in the UK does not ask us to do everything possible to prevent an accident, only what is 'reasonably practicable'.

- **Who should see the report**? In many companies the circulation is kept to a minimum. Very understandably, authors and senior managers do not wish everyone to know about their failures. But this will not prevent the accident happening again. The report should be sent (in an edited form if lengthy) to those people, in the same and other works and departments, who use similar equipment or have similar problems and may be able to learn from the recommendations. In large companies the safety adviser should extract the essential information from the reports he receives and circulate it in a periodic newsletter. My book, *What Went Wrong?*[25] contains many extracts from the monthly *Safety Newsletters* I wrote when I was working for ICI.

Note that in the report on a minor accident in Chapter 15, Appendix the author did not see the deeper layers of the onion but the works manager did, and and asked for further actions.

Many people feel that an accident report is incomplete if it does not recommend a change to the plant, but sometimes altering the hardware will not make another accident less likely. If protective equipment has been neglected, will it help to install more protective equipment? (see Chapter 6).

10   Precept or story?

Western culture, derived from the Greeks, teaches us that stories are trivial light-hearted stuff, suitable for women and children and for occasional relaxation but not to be compared with abstract statements of principles. The highest truths are non-narrative and timeless.

In fact it is the other way round. We learn more from stories, true or fictional, than from statements of principle and exhortations to follow them. Stories describe models which we can follow in our own lives and can help us understand what motivates other people. They instigate action more effectively than codes and standards and have more effect on behaviour. We remember the stories in the Bible, for example, better than all the advice and commandments[26].

Most writing on safety follows the Greek tradition. It sets down principles and guidelines and urges us to follow them. If we read them at all we soon get bored, and soon forget. In contrast, stories, that is, accounts of accidents, can grab our attention, stick in our memories and tell us what we should do to avoid getting into a similar mess.

I am not suggesting that codes and standards are not necessary; obviously they are. Once we see the need to use one, we read it. But only a story will convince us that we need to read it.

The story is not mere packaging, a wrapping to make the principles palatable. The story is the important bit, what really happened. The principles merely sum up the lessons from a number of related stories. You may not agree with the principles but you can't deny the story. We should start with the stories and draw the principles out of them, as I try to do. We should not start with the principles and consider the story in their light.

Of course, we don't always follow the advice, implicit or explicit, in the story. We often think up reasons why our plant is different, why 'it can't happen here'. But we are far more likely to be shocked into action by a narrative than by a code or model procedure.

This then is my justification for describing the accidents in this book. In *What went wrong?*[25] I have described simple incidents, mere anecdotes. The stories in this book are the equivalent of novels but boiled down to the length of short stories.

References

1 *Biblical Archaeological Review*, Vol. 14, No. 2, March/April 1988, p. 21.
2 Neal, W., *With Disastrous Consequences...London Disasters 1830–1917*, Hisarlik Press, London, 1992.
3 Center for Chemical Process Safety, *Guidelines for Investigating Chemical Process Incidents*, American Institute of Chemical Engineers, New York, 1993.
4 Craven, A.D., 'Fire and explosion investigations on chemical plants and oil refineries', in *Safety and Accident Investigations in Chemical Operations*, 2nd edition, edited by H. H. Fawcett and W. S. Wood, Wiley, New York, 1982, p. 659.
5 Carson, P.A., Mumford, C.J. and Ward, R.B, *Loss Prevention Bulletin*, No. 065, Oct. 1985, p 1 and No. 070, August 1986, p. 15.
6 Farmer, D., *Health and Safety at Work*, Vol. 8, No. 11, Nov. 1986, p. 54.
7 Ferry, S.T., *Modern Accident Investigation and Analysis*, 2nd edition, Wiley, New York, 1988.
8 Lynch, M.E., 'How to investigate a plant disaster', in *Fire Protection Manual for Hydrocarbon Processing Plants*, 2nd edition, edited by C. H. Vervalin, Vol. 1, Gulf, Houston Texas, 1985, p. 538.
9 Kletz, T.A., *An Engineer's View of Human Error*, 2nd edition, Institution of Chemical Engineers, Rugby, UK, 1991.
10 *Safety and Health at Work: Report of the Committee 1970–1972* (The Robens Report), Her Majesty's Stationery Office, London, 1972, paragraph 261.
11 *The Leakage of Radioactive Liquor into the Ground, BNFL, Windscale, 15 March 1979*, Her Majesty's Stationery Office, London, 1980, paragraph 51.
12 *Hazard Workshop Modules*, Institution of Chemical Engineers, Rugby, UK, various dates. The subjects covered include plant modifications, fires and explosions, preparation for maintenance, furnace fires and explosions, handling emergencies, human error and learning from accidents.
13 Houston, D.E.L., 'New approaches to the safety problem', in *Major Loss Prevention in the Process Industries*, Symposium Series No. 34, Institution of Chemical Engineers, Rugby, UK, 1971, p. 210.
14 Lees, F.P., *Loss Prevention in the Process Industries*, Butterworths, London, 1980, Vol. 1, Section 2.1 and Vol. 2, Section 27.4.
15 Pope, W.C., 'In case of accident, call the computer', in *Selected Readings in Safety*, edited by J. T. Widner, Academy Press, Macon, Georgia, 1973, p. 295.
16 Ramsey, J.D., 'Identification of contributory factors in occupational injury and illness',

in *Selected Readings in Safety*, edited by J. T. Widner, Academy Press, Macon, Georgia, 1973, p. 328.

17 Kletz, T.A., *Hazop and Hazan – Identifying and Assessing Process Industry Hazards*, 3rd edition, Institution of Chemical Engineers, Rugby, UK, 1992.

18 Lees, F.P., *Loss Prevention in the Process Industries*, Butterworths, London, 1980, Vol. 1, Section 8.9.

19 Knowlton, R.E., *A Manual of Hazard and Operability Studies*, Chemetics International, Vancouver, Canada, 1992.

20 Kletz, T.A., *Plant Design for Safety – A User-Friendly Approach*, Hemisphere, New York, 1991.

21 Lees, F.P., *Loss Prevention in the Process Industries*, Butterworths, London, 1980, Vol. 1, Section 8.1.

22 Kletz, T.A., *Lessons from Disaster – How Organisations have No Memory and Accidents Recur*, Institution of Chemical Engineers, Rugby, UK, 1993, Section 7.4.

23 Tuchman, B., *Practicing History*, Ballantine Books, New York, 1982, p. 23.

24 Dever, W.G., in *The Rise of Ancient Israel*, edited by H. L. Shanks, Biblical Archaeology Society, Washington, DC, 1992, p. 42.

25 Kletz, T.A., *What Went Wrong – Case Histories of Process Plant Disasters*, 2nd edition, Gulf, Houston, Texas, 1988.

26 Cupitt, ·D., *What is a Story?* SCM Press, London, 1991.

Chapter 1

Two simple incidents

Last year, at the Wild Animal Park in Escondido, California, my younger daughter got her first glimpse of a unicorn. She saw it unmistakeably, until the oryx she was looking at turned its head, revealing that, in fact, it had two horns. And in that moment, she learned that the difference between the mundane and the magical is a matter of perspective.

B. Halpern[1] (Figure 1.1)

In the same way, when we look at an accident, we may see technical oversights, hazards that were not seen before or management failings; what we see depends on the way we approach it.

This chapter analyses two simple accidents in order to illustrate the methods of 'layered' accident investigation and to show how much more we can see if we look at the the accidents from different points of view. They also show that we should investigate all accidents, including those that do not result in serious injury or damage, as valuable lessons can be learned from them. 'Near misses', as they are often called are warnings of coming events. We ignore them at our peril, as next time the incidents occur the consequences may be more serious. Engineers who brush aside a small fire as of no consequence are like the girl who said by way of excuse that it was only a small baby. Small fires like small babies grow into bigger ones (see Chapter 18).

1.1 A small fire

A pump had to be removed for repair. The bolts holding it to the connecting pipework were seized and it was decided to burn them off. As the plant handled flammable liquids, the pump was surrounded by temporary sheets of a flame-resistant material and a drain about a metre away was covered with a polyethylene sheet. Sparks burned a hole in this sheet and set fire to the drain. The fire was soon extinguished and no one was hurt. The atmosphere in the drain had been tested with a flammable gas detector two hours before burning started but no gas was detected, probably

Figure 1.1 Unicorn or oryx? What we see depends on the way we look
(Copyright: Bill Clark)

because flammable gas detectors will work only when oxygen is present
and there was too little oxygen below the sheet. It is possible, however,
that conditions changed and flammable vapour appeared in the drain
during the two hours that elapsed before burning started.

First layer recommendations: Preventing the accident

In future we should:

- Cover drains with metal or other flame-resistant sheets before allowing
  welding or burning nearby.
- Test the atmosphere *above* the sheets, not below them.
- Test the atmosphere immediately before welding starts, not several
  hours before. In addition, install a portable flammable gas detector
  which will sound an alarm if conditions change and gas appears while
  welding or burning are in progress.

These recommendations apply widely, not just on the unit where the
fire occurred, so the information should be passed on to other plants.

Second layer recommendations: Avoiding the hazard

Why were the bolts seized? Lubricants which prevent seizure, even at the
high temperatures used in this case, are available. Whose job is it to see
the need for such lubricants and see that they are used?

In an area where flammable liquids or gases are handled seized bolts would normally be cut off rather than burned off. In the present case access was so poor that it was decided to burn them off. Why was access so poor? The normal policy in the company was to build a model of the plant before detailed design is carried out and to review access for maintenance on the model (as well as access for operations, means of escape and many other matters). What went wrong in this case? Was the model review really carried out and were operating and maintenance people present?

Third layer recommendations: Improving the management system

Did the men on the job understand that flammable gas detectors will not detect flammable gas unless it is mixed with air (or oxygen) in the flammable range. Many operators do not understand this limitation of flammable gas detectors. Is this point covered in their training? What is the best way of putting it across so that people will understand and remember?

The plant instructions said that drains must be covered with flame-resistant sheets when welding or burning take place nearby. Over the years everyone had got into the habit of using polyethylene sheets. Did the managers not notice? Or did they notice and turn a blind eye? ('I've got more important things to do than worry about the use of the wrong sort of sheet'.) To prevent the fire, it needed only one manager to keep his eyes open, see that polyethylene sheets were being used, and ask why. On this plant do the managers spend a few hours per day out on the site with their eyes open or do they feel that wandering round the site can be left to the foremen and that their job is to sit in their office thinking about technical problems?

Note that I am using the word 'manager' in the United Kingdom sense of anyone working at the professionally qualified level and that it includes people who in many United States companies would be called supervisors or superintendents.

Some readers may feel that I am making heavy weather of a minor incident but questions such as these are unlikely to be asked unless an incident or series of incidents throw them into focus. Obviously the answers given and the changes made will depend on whether the incident is an isolated one or if other incidents have also drawn attention to weaknesses in training, managerial powers of observation and so on.

The investigating team for an incident such as this would not normally contain any senior managers and we would not expect the unit manager or supervisor to think of all the second and third layer recommendations. But more senior managers should think of them when they read the report. Nor does it take any longer to think of the deeper recommendations as well as the obvious ones. The resource needed is a realisation that such recommendations are possible and necessary, rather than additional time to spend on investigations.

Event	Recommendations for prevention/mitigation
Drain catches fire	
↑ ←	Test immediately before welding starts not 2 hours before. Use portable gas detector alarms during welding.
Hole burnt in sheet by welding sparks	
↑ ←	Cover drains with metal or other flame-resistant sheets.
Drain tested. No flammable gas detected	
↑ ←	Test above sheet, not below. **Train operators in limitations of gas detectors (& other equipment).**
Drain covered with plastic sheet	
↑ ←	**Regular audits and keeping eyes open might have shown that the wrong sheets were regularly used.**
Decision made to burn off bolts	
↑ ←	*Provide better access so that bolts can be cut off.* *During design, operating staff should review model.*
Pump bolts seized	
↑ ←	*Use high temperature lubricants.*
Ordinary type *Italics* **Bold type**	1st layer: Immediate technical recommendations *2nd layer: Avoiding the hazard* **3rd layer: Improving the management system**

Figure 1.2 Summary of Section 1.1 - A small fire

Figure 1.2 summarises, on a time scale, the events leading up to the accident and the recommendations made. It should be read from the bottom up. First, second and third layer recommendations are indicated by different typefaces. First layer recommendations, the immediate technical ones, are printed in ordinary type, second layer recommendations, ways of avoiding the hazard, are printed in italics and third layer recommendations, ways of improving the management system, are printed in bold type. The same scheme is followed in later chapters though the allocation between categories is inevitably in some cases a matter of opinion. Thus hazop is shown as a means of avoiding the hazard but might equally well be considered a means of improving the management system.

The diagram shows us that there were many opportunities of preventing the accident, by breaking the chain of events that lead up to it at different points. Some of the actions had to be taken shortly before the accident occurred, others a long time before. Some of these actions would have removed the immediate causes while others would have removed the hazard or dealt with the weaknesses in the management system which were the underlying causes.

In general, the most effective actions are those at the bottom of the diagrams. If we are constructing defences in depth we should make sure that the outer defences are sound as well as the inner ones. Protective measures should come at the bottom of the accident chain and not just at the top. In many of the accidents described later there was too much dependence on the inner defences, the protective measures at the top of the accident chain. When these defences failed there was nothing in reserve.

1.2   A mechanical accident

This section describes an accident to a mixer – but it is really about all accidents, so please read it even if you never have to design or operate a mixer.

A mixing vessel of 1 m³ (264 US gallons) capacity was fitted with a hinged, counter-weighted lid (Figure 1.3). To empty the vessel the lid was opened (Figure 1.4), the vessel rotated anti-clockwise and the contents shovelled out (Figure 1.5). One day the lid fell off and hit the man who was emptying the vessel. Fortunately his injuries were not serious.

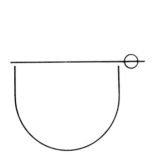

Figure 1.3 The mixing vessel in use

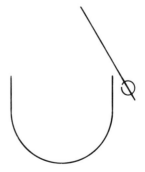

Figure 1.4 The lid is opened

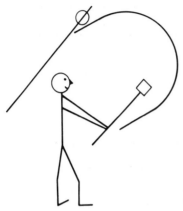

Figure 1.5 The vessel is rotated so that the contents can be removed

It was then found that the welds between the lid and its hinges had cracked. It was a fatigue failure, caused by the strains set up by repeated opening and closing of the lid. There was nothing wrong with the original design but the lid had been modified about ten years before the incident occurred and, in addition, some repairs carried out a few years before had not been to a high enough standard.

Detailed recommendations were made for the repair of the lid. Though necessary they do not go far enough. If we look at the inner layers of the onion, four more recommendations are seen to be necessary (Figure 1.7):

Figure 1.6

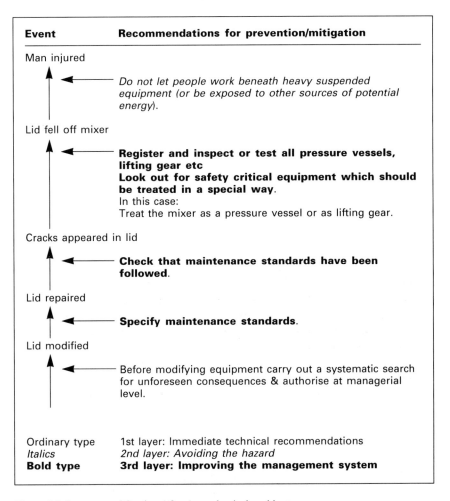

Figure 1.7 Summary of Section 1.2 - A mechanical accident

(1) What is the system for the control of modifications? Is anyone who thinks he can improve a piece of equipment allowed to do so? Before any equipment is modified the change should be approved by a professionally qualified engineer who tries to make sure that the change is to the same standard as the original design and that there are no unforeseen side-effects (see Section 7.1). This is one of the lessons of Flixborough (Chapter 8). Many other accidents have occurred because plants or processes were modified and no one foresaw the consequences of the change[2-4].

After a modification has been made the engineer who approved it should inspect the completed work to make sure that his intentions have been followed and that the modification looks right. What does not look right is usually not right and should at least be checked (Figure 1.6).

(2) Why were the repairs not carried out to a high enough standard? Who is (or should be made) responsible for specifying the standard of repairs and modifications and checking that work has been carried out to this standard? Does anyone know the original design standard?

(3) Cracks would have been present in the welds for some time before they failed completely and could have been detected if the lid had been inspected regularly. The company concerned registered and inspected all pressure vessels and, under a separate scheme, all lifting gear. However, the mixer was not registered under either scheme as it operated at atmospheric pressure and so was not a pressure vessel and it was not recognised as lifting gear. Yet its failure could be as dangerous as the failure of vessels or lifting gear. It should be registered under one of the schemes. It does not matter which, provided the points to be looked for during inspection are noted.

Many other accidents have occurred because equipment was not recognised as coming into one of the categories that should be registered and inspected or treated in some special way. Chapter 7 discusses an accident that occurred because the size of an open vent was reduced without checking that the smaller size would be adequate. No one realised that the vent was the vessel's relief valve and should be treated like a relief valve: its size should not be changed unless we have gone through the same procedure as we would go through before changing the size of a relief valve.

Similarly, if a relief valve has been sized on the assumption that a non-return (check) valve (or two in series) will operate, the non-return valve should be included in the register of relief valves and inspected regularly, say, once per year. If a relief valve has been sized on the assumption that a control valve trim is a certain size, this control valve should be included in the relief valve register, its size should not be changed without checking that the new size will be adequate and the valve should be scheduled for regular examination, say, once per year, to check that the original trim is still in position. The control valve register should be marked to show that this valve is special.

(4) People should not normally be expected to work underneath heavy suspended objects. This was apparently not known to those who designed, ordered and operated the mixer though as far back as 1891 the House of Lords (in *Smith v Baker & Sons*) ruled that it was an unsafe system of work to permit a crane to swing heavy stones over the heads of men working below.[5] The company carried out regular safety audits but though the mixer had been in use for ten years no one recognised the hazard. What could be done to improve the audits? Perhaps if outsiders had been included in the audit teams they would have picked up the hazard.

In Japan in 1991 fourteen people were killed and nine were seriously injured when a steel girder, part of a new railway line, fell onto a row of cars. The girder was 63 m long, weighed 53 tonnes and was supported on eight jacks[6].

Just as people should not work below equipment which is liable to fall, so they should not work above equipment which is liable to move upwards. At an aircraft factory a man was working above a fighter plane which was nearly complete. The ejector seat went off and the man was killed. In general potential energy and trapped mechanical energy are as dangerous as trapped pressure and should be treated with the same respect. Before working on a fork lift truck or any other mechanical handling equipment we should make sure that it is in the lowest energy state, that is, in a position in which it is least likely to move as it is being dismantled. If equipment contains springs, they should be released from compression (or extension) before the equipment is dismantled.

These facts show that thorough consideration of a simple accident can set in motion a train of thought that can lead to a fresh look at the way a host of operations are carried out.

References

The two incidents described in this chapter originally appeared, in a much shorter form, in *Health and Safety at Work*, Vol. 7, No. 1, Jan. 1985, p, 8 and *Occupational Safety and Health*, Vol. 15, No. 2, Feb. 1985, p. 25.

1   Halpern, B., in *The Rise of Ancient Israel*, edited by H. L. Shanks, Biblical Archaeology Society, Washington, DC, 1992, p. 105.
2   Kletz, T. A., *What Went Wrong? Case Histories of Process Plant Disasters*, 2nd edition, Gulf Publishing Co, Houston, Texas, 1988, Chapter 2.
3   Lees, F. P., *Loss Prevention in the Process Industries*, Butterworths, London, 1980, Vol. 2, Chapter 21.
4   Sanders, R. E., *Management of Change in Chemical Plants: Learning from Case Histories*, Butterworth-Heinemann, London, 1993.
5   Farmer, D., *Health and Safety at Work*, Vol. 8, No. 11, Nov. 1986, p. 61.
6   Yasuda Fire and Marine Insurance Co, *Safety Engineering News* (Japan), No. 17, April 1992, p. 7.

Chapter 2

Protective system failure

. . .although fragmentation is not actually a problem, the cures for it can be.

A. Solomon, *Daily Telegraph,* 27 February 1989, p. 27.

To meet a demand from some customers for a product containing less water, a small drying unit was added to a plant which manufactured an organic solvent. The solvent, which was miscible with water, was passed over a drying agent for about eight hours; the solvent was then blown out of the the drier with nitrogen and the drier regenerated. There were two driers, one working, one regenerating (Figure 2.1).

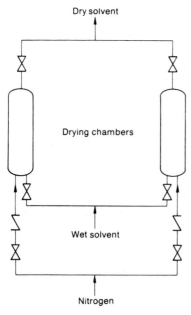

Dry solvent

Drying chambers

Wet solvent

Nitrogen

Figure 2.1 Drying unit in which the accident occurred. (Regeneration lines not shown)

As the drying unit was some distance from the control room the instruments associated with it were mounted on the outdoor control panel shown in Figure 2.2. The top half of the panel contained pneumatic instruments, the lower half electrical equipment associated with the change-over of the driers. The control panel was located in a Zone (Division) 2 area, that is, an area in which a flammable mixture is not likely to occur in normal operation and, if it does occur, will exist for only a short time (say, for a total of not more than ten hours per year). The electrical equipment could not, at the time of construction, be obtained in a flameproof or non-sparking form suitable for use in a Zone 2 area. It was therefore mounted in a metal cabinet, made from thin metal sheet, which was continuously purged with nitrogen. The nitrogen was intended to keep out any solvent vapour that might leak from the drying unit or the main plant. Such leaks were unlikely, and if they did occur, would probably be short-lived, but the Zone 2 classification showed that they could not be ruled out. A pressure switch isolated the electricity supply if the pressure in the cabinet fell below a preset value, originally 1/2 inch water gauge (0.125 kPa).

No solvent or other process material was connected to the control panel.

Figure 2.2 Instruments controlling the drying unit were located in this outdoor panel. Electrical equipment was purged with nitrogen

Despite these precautions an explosion occurred during the commissioning of the drying unit. It had been shut down for a few days and was ready to restart. A young graduate had been given the job of commissioning the unit as his first industrial experience. Standing in the position shown in Figure 2.2 he personally switched on the electricity supply. There was an explosion and the front cover was blown off the metal cabinet, hitting him in the legs. Fortunately no bones were broken and he returned to work after a few days.

For an explosion we need fuel, air (or oxygen) and a source of ignition and we shall consider these separately before looking at the underlying factors.

2.1   The fuel

There was no leak from the drying unit or the main plant at the time and there was no flammable vapour present in the atmosphere. The fuel did not leak into the metal cabinet from outside, the route which had been foreseen and against which precautions had been taken, but entered with the nitrogen. The nitrogen supply was permanently connected to the driers by single isolation valves and non-return (check) valves as shown in Figure 2.1. The gauge pressure of the nitrogen was nominally 40 psi (almost 3 bar) but fell when the demand was high. The gauge pressure in the driers was about 30 psi (2 bar). Solvent therefore entered the nitrogen lines through

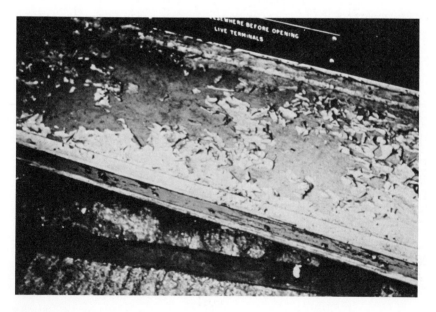

Figure 2.3 This view of the cabinet's inside shows paint attacked by solvent, suggesting that vapour had been getting in for some time

leaking valves and found its way into the inside of the cabinet. The solvent had to pass through a non-return (check) valve but these valves are intended to prevent gross back-flow not small leaks. In the photograph of the inside of the cabinet (Figure 2.3), taken immediately after the explosion, the damaged paintwork shows that solvent must have been present for some time. However, solvent vapour and nitrogen will not explode and the solvent alone could not produce an explosive atmosphere.

2.2 The air

Air diffused into the cabinet as the nitrogen pressure had fallen to zero for some hours immediately before the accident. The unit was at the end of the nitrogen distribution network and suffered more than most units from deficiencies in the supply. It is difficult to get airtight joints in a cabinet made from thin metal sheets bolted together and air diffused in through the joints. The solvent may have affected the gaskets in the joints and made them more porous.

2.3 The source of ignition

The source of ignition was clearly electrical as the explosion occurred when the electricity was switched on. However, the low-pressure switch

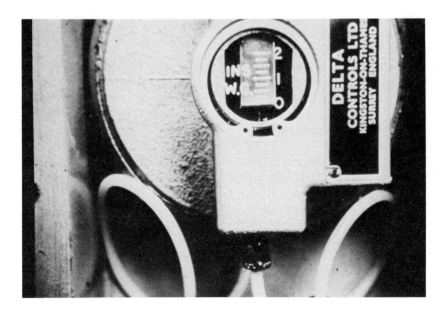

Figure 2.4 Note that on the switch, as shown here with the cover removed, the set-point has been reduced enough to disarm the protective equipment

should have isolated the supply. The reason it did not do so is shown by Figure 2.4, a photograph of the pressure switch with the cover removed. It will be seen that the set-point has been reduced from 1/2 inch water gauge to zero. The switch cannot operate unless the pressure in the cabinet falls below zero, an impossible situation. The protective equipment had been effectively disarmed (that is, made inoperable).

The switch was normally covered by a metal cover and the set-point was not visible. Only electricians were authorised to remove the cover.

2.4   First layer recommendations

The following recommendations were made during the enquiry immediately following the incident:

The fuel To prevent contamination of the nitrogen it should not be permanently connected to the driers by single valves but by hoses which are disconnected when not in use or by double block and bleed valves. In addition, in case the nitrogen pressure falls while the nitrogen is in use, there should be a low pressure alarm on the nitrogen supply set a little above the pressure in the driers.

The first recommendation applies whenever service lines have to be connected to process equipment, and the second one applies whenever the pressure in a service line is liable to fall below the process pressure. (If the process pressure is liable to rise above the service pressure, there should be a high pressure alarm on the process line.) Neglect of these precautions has resulted in nitrogen leaks catching fire, air lines setting solid and steam lines freezing.

The accident at Three Mile Island (Chapter 11) and the incident described in Section 5.6 were also initiated by backflow into service lines.

In the longer term a more reliable nitrogen supply should be provided, either by improving the supply to the whole plant or by providing an independent supply to equipment which is dependent on nitrogen for its safety.

The air It is impossible to make an airtight box from thin metal sheets bolted together. If the nitrogen supply could not be made more reliable then the metal cabinet should have been made more substantial.

The source of ignition Alterations in the set-points of trips (and alarms) should be made only after authorisation in writing at managerial level. They should be recorded and made known to the operators.

Set-points should be visible to the operators; the pressure switch should therefore have a glass or plastic cover. Unfortunately, carrying out this recommendation is not as easy as it sounds. The switch was a flameproof one and could not be modified without invalidating its certification. Redesign had to be discussed with and agreed by the manufacturer and followed by recertification.

All trips (and alarms) should be tested regularly. This was the practice on the plant concerned but as the drying unit was new it had not yet been added to the test schedules. Obviously new equipment (of any sort) should be scheduled for whatever testing and inspection is considered necessary as soon as it is brought into use.

These recommendations also apply to all plants.

2.5   Second layer recommendations

After the dust had settled and those concerned had had time to reflect, they asked why the trip had been disarmed. It seemed that the operators had had difficulty maintaining a pressure of 1/2 inch water gauge in the leaking cabinet. The trip kept operating and shutting down the drying unit. They complained to the electrical department who reduced the set-point to 1/4 inch water gauge. This did not cure the problem. Finally one electrician solved the problem by reducing the set-point to zero. He did not tell anyone what he had done and the operators decided he was a good electrician who had succeeded where the others had failed. After the explosion he chose anonymity.

The designers had not realised how difficult it is to maintain even a slight pressure in a cabinet of thin metal sheets. If they had done so they might have installed a low flow alarm instead of a low pressure alarm. In addition they did not know that the nitrogen supply was so unreliable. The plant data sheets showed that a nitrogen supply at a gauge pressure of 40 psi (almost 3 bar) was available and they took the data sheets at their word. If a hazard and operability study had been carried out on the design with the unit manager present then this would probably have come to light. A hazard and operability study was carried out but only on the process lines, not on the service lines. Many other incidents have shown that it is necessary to study service lines as well as process lines.[1]

2.6   Third layer recommendations

Further recommendations were made when the explosion was selected for discussion by groups of managers and designers as described in Part 4 of the Introduction. (Some of these recommendations deal with ways of avoiding the hazard and have therefore been classified as second layer in Figure 2.5.)

The cabinet could be pressurised with air instead of nitrogen. The purpose of the nitrogen was to prevent solvent vapour diffusing in from outside. Air could do this equally well and the reliability of the compressed air supply was much better than that of the nitrogen supply. Compressed air was also much cheaper.

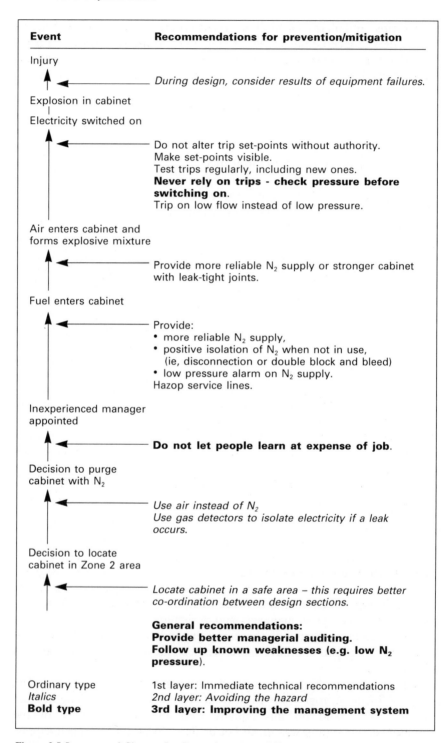

Figure 2.5 Summary of Chapter 2 – Protective system failure

Compressed air has another advantage. If anyone had put his head in the cabinet to inspect or maintain the contents, without first making sure that the nitrogen supply was disconnected, he could have been asphyxiated. Compressed air would have removed this hazard. Nitrogen is widely used to prevent fires and explosions but many people have been killed or overcome by it. It is the most dangerous gas in common use.[2]

The safety equipment, installed to guard against a rather unlikely hazard – a leak of solvent near the control panel – actually produced a greater hazard. Is there, then, another way of guarding against the original hazard? One possibility is installing flammable gas detectors to detect a leak of solvent and then using the signal from them to isolate the electricity supply.

Did the control panel have to be in a Division 2 area? It did not. It was put in what was believed to be a convenient location without considering the electrical classification of the area. The electrical design engineer then asked for nitrogen purging so that he could meet the requirements of the classification. He did not ask if the control panel could be moved. That was not his job. His job was to supply equipment suitable for the agreed classification. It was no one's job to ask if it would be possible to change the classification by moving the equipment. In fact, if the control panel had been moved a few metres it would have been in a safe area.

This illustrates our usual approach to safety. When we recognise a hazard – in this case a leak of solvent vapour ignited by electrical equipment – we *add on* protective equipment to control it. We rarely ask if the hazard can be removed.

The man who was injured was, as already stated, a young inexperienced graduate. A more experienced man might have foreseen the hazards and taken extra precautions. Thus the nitrogen pressure should have been checked before the electricity was switched on. It is bad practice to assume that a trip will always work and rely on it. Letting people learn by doing a job and making mistakes is excellent training for them (though not always good for the job) but is not a suitable procedure when hazardous materials are handled. If the young engineer had received some training in loss prevention during his university course, or in a company training scheme, he might have been better prepared for the task he was given. Today, in the UK, though not in most other countries, all undergraduate chemical engineers get some training in loss prevention.[3]

2.7   Other points

At first sight this explosion seemed to result from the coincidence of four events, three of them unlikely:

(1) A low nitrogen pressure, which allowed solvent to contaminate the nitrogen, followed by

(2) a complete nitrogen failure, which allowed air to diffuse into the cabinet;

(3) disarming of the trip which should have isolated the electricity when the pressure in the cabinet was low and

(4) the triggering event, someone switching on the electricity.

In fact the first three were not events occurring at a point in time but unrevealed or latent faults[4] that had existed for long periods; the first had existed on and off for weeks, the second for several hours, from time to time, and the third for days or weeks. It was therefore almost inevitable that sooner or later the triggering event – a true event – would coincide with the three unrevealed or latent faults and an explosion would result.

Accidents are often said to be due to unlikely coincidences, thus implying that people could not reasonably have been expected to foresee them and take precautions. When the facts are established, however, it is usually found that all but one of the so-called events were unrevealed faults that had existed for some time. When the final event occurred the accident was inevitable.

While each error that occurred might have been excused, taken as a whole they indicate poor methods of working and a lack of managerial auditing, by no means unusual at the time the accident occurred. In particular:

• The unsatisfactory state of the nitrogen supply was known and improvements were being studied but perhaps not with sufficient urgency. In the meantime more thought might have been given to the prevention of contamination.

• There should have been better control of trip set-points.

The flimsy box containing the electrical equipment was connected to a source of pressure. If the box was overpressured for any reason the weakest part, the bolted front, would blow off and injure anyone standing in front of the instrument panel. It would have been better to place the cover on the far side of the box, away from the operator. Designers should always consider the results of equipment failure.

Similarly, horizontal cylindrical pressure vessels should be located so that if they rupture, for example, as the result of exposure to fire, the bits do not fly in the direction of occupied buildings or other concentrations of people (see also Section 7.4).

References

This chapter is based on an article which was published in *Hydrocarbon Processing*, Vol. 59, No. 11, Nov. 1979, p. 373 and thanks are due to Gulf Publishing Co. for permission to quote from it.

1 Kletz, T.A.,*What Went Wrong? Case Histories of Process Plant Disasters*, 2nd edition, Gulf Publishing Co, Houston, Texas, 1988, Section 18.2.

2 Kletz,T.A., *What Went Wrong? Case Histories of Process Plant Disasters*, 2nd edition, Gulf Publishing Co.., Houston, Texas, 1988, Section 12.3.
3 Kletz,T.A., *Plant/Operations Progress*, Vol. 7, No. 2, April 1988, p. 95.
4 Wreathall,J., 'Human error and process safety', *Health, Safety and Loss Prevention in the Oil, Chemical and Process Industries*, Butterworth-Heinemann, Oxford, 1993, p. 82.

Chapter 3

Poor procedures and poor management

Prudence dictated that Scott plan for a wide margin of safety. He left none and thereby killed not only himself but four others.

J. Diamond[1]

A crude oil distillation unit was being started up after a major turnaround. It had taken longer than expected so stocks of product were low and it was important to get the unit on-line as soon as possible. The manager (the supervisor in most US companies) therefore decided to be present throughout the night so that he could deal promptly with any problems that arose. Perhaps also his presence might discourage delay. He was a young graduate who had been in the job for only a year and an additional reason for being present was to see for himself what happened during a major start-up.

The distillation column was warming up. It had been washed out with water before the shut-down and the water left in the column had distilled into the reflux drum and had half filled it. There was a layer of light oil containing some liquefied petroleum gas (LPG) on top of the water. (Some water was always produced but as the column had been washed out the production of water was greater than usual.) Two pumps were connected to the reflux drum as shown in Figure 3.1. The water pump took suction from the bottom of the drum and sent the water to a sour water scrubber for purification and discharge to drain; the oil pump took suction from a point about 0.3 m (1 foot) above the bottom and provided reflux and product take-off. Neither pump had been started up.

The foreman asked an operator to start up the water pump. He discovered that a slip-plate (spade) had been left in the suction line to the pump on the drum side of the isolation valve (Figure 3.1). All the branches on the drum had been slip-plated during the turnaround to isolate the drum for entry. The other slip-plates had been removed but this one had been overlooked by the fitter who removed them and this was not noticed by the process foreman when he accepted back the permit-to-work.

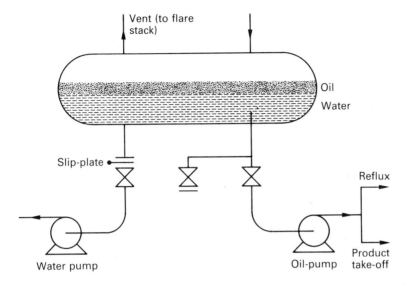

Vent (to flare stack)

Oil
Water

Slip-plate

Reflux

Water pump

Oil-pump   Product take-off

Figure 3.1 The reflux drum and connections, showing the position of the slip-plate

The manager estimated that shutting down the reboiler furnace, allowing it to cool, fitting a hose to the spare branch on the reflux drum, draining the contents to a safe place, removing the slip-plate and warming up again would result in twenty-four hours delay. The maintenance foreman, a man of great experience, who was also present, offered to break the joint, remove the slip-plate and remake the joint while the water ran out of it. He could do it, he said, before all the water ran out and was followed by the oil; he had done such jobs before.

After some hesitation the manager agreed to let the foreman go ahead. He dressed up in waterproof clothing and watched by the process team, unbolted the joint and removed the slip-plate while the water sprayed out. Unfortunately he tore one of the compressed asbestos fibre gaskets, half of it sticking to one of the joint faces. Before he could remove it and replace it, all the water ran out and was followed by the oil.

Some of the LPG flashed as the oil came out of the broken joint. This cooled the joint and some ice formed, making it impossible to remake the joint. The foreman abandoned the attempt to do so.

The reboiler furnace was only 30 m away. As soon as the oil appeared, one of the process team pressed the button which should have shut down the burners. Nothing happened. The process team had to isolate the burners one by one while the oil and vapour were spreading across the level ground towards the furnace. Fortunately, they did so without the vapour igniting.

Afterwards it was discovered that the protective system on the furnace had given trouble a day or two before the turnaround started. The process foreman on duty therefore took a considered decision to by-pass it until the turnaround, when it could be repaired. Unfortunately there was so much work to be done during the turnaround that this late addition to the job list was overlooked.

Although there was no injury or damage, both could easily have occurred, and there was a day's loss of production. The incident was therefore thoroughly investigated, as all dangerous occurrences and near-misses should be. In the following we assume that each member of the five-man investigating panel wrote a separate report, each emphasising different causes and making different recommendations. Of course, it was not really as tidy as this. The first two sets of recommendations were made at the time. The others were made later when the incident was selected for discussion in the manner described in Part 4 of the Introduction.

The five sets of recommendations are not alternatives. All are necessary if a repeat of the accident is to be prevented.

3.1   Report 1 – The slip-plate is the key

The author of this report pointed out that the incident was entirely due to the failure to remove the slip-plate before the start-up. If the slip-plate had been removed the incident would not have occurred. He made the following recommendations which, like all those made in the other reports, apply generally and not just on the plant where the incident occurred:

- All slip-plates should be listed on a permit-to-work. It is not sufficient to say 'Slip-plate (or de-slip-plate) all branches on reflux drum'. Instead their positions should be listed and identified by numbered tags. 'All' is a word that should be avoided when writing instructions on safety matters. Someone asked to slip-plate 'all lines' does not know whether he has to slip-plate two, three, four or many.
- When a maintenance job is complete the process foreman, before accepting back the permit-to-work, should check that the job done is the job he wanted done and that all parts of it have been completed.
- In addition, all slip-plates inserted during a shut-down should be entered on a master list and a final inspection made, using this list, before start-up.
- If the slip-plate had been inserted below the isolation valve it would have been possible to remove it with the plant on-line. Nevertheless we should continue to insert slip-plates on the vessel side of isolation valves, as if they are fitted on the far side, liquid might be trapped between the slip-plate and a closed valve and then slowly evaporate while people were working in the vessel. Such incidents have occurred.

Chapter 5 describes an accident which was the result of failure to *insert* a slip-plate.

If fibreglass reinforced plastic lines, used to handle acids, have to be slip-plated it is common practice to use a steel slip-plate covered by a disc of polytetrafluoroethylene (PTFE). Sanders[2] has described an incident in which such a slip-plate was removed from the overflow line in an acid tank but the PTFE disc was left behind. Nothing happened for over a year, until the water in a small scrubber in the vent line froze during a cold spell; the tank was then overpressured during filling and blew apart.

3.2   Report 2 – Better control of protective systems

The author saw the failure to shut down the furnace as the key point. Emergencies of one sort or another are inevitable from time to time and we must be able to rely on our protective equipment. He recommended that:

- Protective equipment should not be by-passed or isolated unless this has been authorised in writing by a responsible person.
- If it is by-passed or isolated this should be signalled to the operators in some way, for example by a light on the panel. A note in the shift log is not enough.
- All trips should be tested after a major turnaround and all trips that have been repaired or overhauled should be tested before they are put back into service. If testing is necessary outside normal hours and cannot be carried out by the shift team then the people who carry out the testing should be called into work. (Chapter 2 described another incident which occurred because a trip had been made inoperable.)

3.3   Report 3 – Don't rush

The author saw the key event as a rushed decision by the manager. Few problems on a large plant are so urgent that we cannot delay action for 15 minutes while we talk them over. If those concerned had paused for a cup of tea they would have realised that removing the slip-plate was more hazardous than it seemed at first sight and that there were other ways of avoiding a shut-down. I cannot vouch for the authenticity of Figure 3.2 but I agree with the message (compare Section 11.7).

Removing the slip-plate was more hazardous than it seemed at first sight because the gauge pressure at the slip-plate, due to the head of liquid, was nearly 10 psi (0.7 bar) higher than the pressure in the reflux drum (a gauge pressure of about 15 psi or 1 bar). This was not realised at the time. It might have been realised if those present had given themselves time to talk over the proposed course of action.

It is a useful rule of thumb to remember that if a column of liquid in a pipeline is x feet tall, it will spray out x feet horizontally if a flanged joint is slackened.

Figure 3.2 The notice may not be authentic but the message is sound

A shutdown could have been avoided with less risk by freezing the water above the slip-plate with solid carbon dioxide (dry ice) or by injecting water into the reflux drum via the spare branch shown in Figure 3.1 so as to maintain the level. Another possible way of avoiding the shutdown would be to remove the pump, pass a drill through the valve and drill through the slip-plate. This method could, of course, only be used if the valve was a straight-through type.

The company manufactured solid carbon dioxide at another site 10 km away and it had been used before to freeze water lines in the plant, but in the heat of the moment no one remembered this.

As a general rule, when we have to decide between two courses of action, both of which have disadvantages, there are often alternative actions available which we have not thought of.

3.4   Report 4 – Who was in charge?

The author of this report saw the accident as due to the failure of the young manager to stand up to the maintenance foremen. The manager's situation was difficult. The foreman was a strong personality, widely respected as an experienced craftsman, old enough to be the manager's father, and he assured the manager that he had done similar jobs before. It was 3 am, not the best time of day for decisions. The manager could not be blamed. Nevertheless sooner or later every manager has to learn to stand up to his foremen, not disregarding their advice, but weighing it in the balance. He should be very reluctant to overrule them if they are advocating caution, more willing to do so if, as in this case, they advocate taking a chance.

The maintenance foreman felt partly responsible for the non-removal of the slip-plate. This made him more willing than he might otherwise have been to compensate for his mistake by taking a chance. A more experienced manager would have realised this.

In all walks of life, leaders have to decide whether or not to follow the advice, often conflicting, that they receive from colleagues, subordinates and experts. It now seems that one of the reasons for the failure of Robert Scott's 1911 expedition to the South Pole was his unwillingness to listen to his companions' views[1,3]. Commenting on this, Diamond writes:

> So when a companion tells me that my proposal is dangerous, I have to be alert; he may be right, or he may just be lazy. The only universal rule I have for such situations is at least to listen to advice and evaluate it.

3.5   Report 5 – The climate in the works

This author went deeper than the other two. He saw the incident as due to a failure to give sufficient emphasis to safety throughout the organisation. What would the works manager have said the next morning if he found that the start-up had been delayed? Would he have commented first on the low stocks and lost production or would he have said that despite the low stocks he was pleased that no chances had been taken?

The young manager was not working in a vacuum. His judgement was influenced by his assessment of his bosses' reactions and by the attitude to safety in the company, as demonstrated by the actions taken or remarks made in other situations. Official statements of policy have little influence. We judge people by what they do, not what they say. If anyone is to be blamed, it is the works manager for setting a climate in which his staff felt that risk-taking was legitimate.

Did the manager feel that he had been given, by implication, contradictory instructions (like the operators at Chernobyl; see Chapter 12): in this case to get the plant back on line as soon as possible and, at the same time, to follow normal safety procedures? Junior managers and foremen often find themselves in this position. Senior managers stress the importance of output or efficiency but do not mention safety. So their subordinates assume that safety takes second place. They are in a 'no-win' situation. If there is an accident they are blamed for not following the safety procedures. If the required output or efficiency are not achieved they are blamed for that. Managers, when talking about output and efficiency, should bring safety into the conversation. **What we don't say is as important as what we do say**.

It may be right, on occasions to relax the safety rules, but if so this should be clearly stated, not hinted at.

How, if at all, did the young manager's training in the company and at University prepare him for the situation in which he found himself? Probably not at all. Today, in the UK, all undergraduate chemical engineers receive some training in loss prevention though it is unlikely to cover situations such as that described.

3.6   Other comments

Was it wise to attempt to shut down the furnace, either automatically or manually? It is probably safer to keep a furnace operating if there is a leak of flammable gas or vapour nearby. The flame speed is much less than that of the air entering a furnace (unless hydrogen is present) so flashback from the furnace is very unlikely. On the other hand, if the furnace is shut down, the gas or vapour may be ignited by hot brickwork.

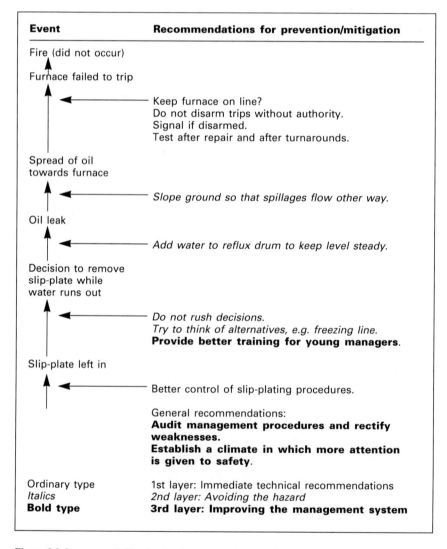

Figure 3.3 Summary of Chapter 3 – Poor procedures and poor management

The ground should have been sloped so that any liquid spillage flowed away from the furnace. In general, spillages should flow away from equipment, not towards it (see Chapter 5).

The incident illustrates the comment in Part 3 of the Introduction that accidents are rarely the fault of a single person but that responsibility is spread in varying degrees amongst many people: those who failed to remove the slip-plate, those who by-passed the furnace trip and then failed to make sure that it was repaired, the young manager, those responsible for his training and guidance, the maintenance foreman, the works manager. Any of these, by doing their job better, had the power to prevent the incident. But the systems for controlling slip-plating and the disarming of trips were unsatisfactory to say the least. At the operating level, those concerned were following custom and practice, and the greater responsibility is therefore that of the works manager and his senior colleagues who either failed to recognise the deficiencies in their procedures or failed to do anything about them.

References

This chapter is based on a paper which was published in *Plant/Operations Progress*, Vol. 3, No. 1, January 1984, p. 1, and thanks are due to the American Institute of Chemical Engineers for permission to quote from it.

1   Diamond, J., *Discover*, Vol. 10, No. 4, April 1989, p. 75.
2   Sanders, R. E., *Management of Change in Chemical Plants: Learning from Case Histories*, Butterworth-Heinemann, Oxford, 1993.
3   Griffiths, T., *Judgement over the Dead*, Verso, Oxford, 1986.

Chapter 4

A gas leak and explosion – The hazards of insularity

It is hubris to imagine that we can infallibly prevent a thermodynamically favoured event.

P. G. Urben, reviewing the first edition of this book[1]

Four men were killed, several injured and a compressor house was destroyed when a leak of ethylene ignited. Figure 4.1 shows some of the damage. Explosion experts estimated that between 5 and 50 kg of ethylene leaked out into the building during the eight minutes or so that elapsed between the start of the leak and the explosion. Detailed examination of the wreckage enabled the source of the leak to be identified. It was a badly made joint on a small bore line.

The recommendations made to prevent similar explosions fell into four groups:

- Ways of preventing leaks or making them less likely.
- Ways of minimising their effects.
- Ways of reducing the probability of ignition.
- Underlying all these, ways of removing the managerial ignorance and incorrect beliefs that led to the accident.

Most of the recommendations apply to other plants handling hazardous materials, particularly flammable liquids and gases, and many apply to all process plants.

4.1 Preventing leaks

As already stated the leak of ethylene occurred from a badly made joint on a small diameter line, about 1/2 inch (12 mm) internal diameter. The joint was a special type, suitable for use at high pressures and assembling it required more skill than is required for an ordinary flanged joint. Once the joint had been made it was impossible to see whether or not it had been made correctly. It would be easy to say that more care should be

Figure 4.1 Some of the damage caused by the explosion

taken in joint making but this temptation was resisted and the underlying reasons why the joint had been badly made were investigated. There were two reasons.

(1) At one time the assembly of high pressure joints was carried out only by a handful of craftsman who were specially trained and judged to have the necessary level of skill and commitment. This was resented by the other craftsmen and a change was made, all craftsmen being trained to the necessary standard, or so it was thought, and the work carried out by whoever was available. Manpower utilisation was improved. Unfortunately some of the newly trained men did not have the necessary skill, or perhaps did not understand the importance of using their skill to the full, and the standard of joint-making deteriorated. Following the explosion a return was made to the original system.

In addition, several other actions were taken:

(a) A standard for the quality of joints was specified.

(b) An attempt was made to explain to the craftsmen why such a high quality of joint making was necessary and the results that would follow if joint making was not up to standard. Obviously this task was made easier by the results of the explosion, which were known to all concerned, and it would have been more difficult to do the same on a plant which had not suffered a serious leak. Nevertheless it is something that should be attempted on many plants.

(c) A system of inspection was set up. The joints had to be inspected after the surfaces had been prepared and before the joint was made. Initially all joints were inspected but as the inspectors gained confidence in the ability of individual craftsmen they inspected only a proportion, selected at random. Obviously, in the atmosphere that existed after the explosion, the craftsmen accepted inspection more readily than they would have done at other times.

(d) Better tools, to make joint making easier, were developed and made available.

These actions reduced the leak frequency by a factor of about twenty.

(2) Although leaks were quite common before the explosion, about ten per month, mostly very small, nobody worried about them. The attitude was, 'They can't ignite because we have eliminated all sources of ignition'. Unfortunately this view was incorrect. It is almost impossible to eliminate all sources of ignition and, as we shall see later, not everything that could have been done to eliminate them had in fact been done. A study made after the explosion showed that several ignitions had occurred in the past, on this and similar plants, but that nevertheless the probability of ignition was very low, only about one leak in ten thousand igniting. This low probability may have been due to the fact that the high pressure gas dispersed by jet mixing. On most plants the probability of ignition of small leaks is probably between one in ten and one in a hundred. (For large leaks, greater than several tonnes, the probability of ignition is much higher, greater than one in ten and perhaps as high as one in two.)

A third reason for the large number of leaks was the large number of joints and valves. The plant consisted of several parallel streams each containing three main items of equipment. Their reliability was not high and so, in order to maintain production, a vast number of crossovers and isolation valves was installed so that any item could be used in any stream. The money spent in providing all this flexibility might have been better spent in investigating the reasons why on-line time was so poor. Later plants, built after the explosion, had fewer streams and fewer crossovers.

4.2   Minimising the effects of leaks

Although much can be done, and was done after the explosion, to make leaks less probable, they were still liable to occur from time to time and several actions were taken, to various extents, to minimise their effects. Taken together they constitute a defence in depth. If one line of defence fails, the next one comes into operation. They may be summarised by the words: DETECT, WARN, ISOLATE, DISPERSE,VENT.

Detect Flammable gas detectors provide a cheap and effective means of giving early warning that a leak has occurred. Action can then be taken to isolate the leak, evacuate personnel, call the fire service in case the leak ignites, and so on. They should be installed on all plants where experience shows that leaks of hazardous materials are liable to occur, such as all plants handling liquefied flammable gases. It is particularly important to install them in buildings as very small leaks indoors can cause serious damage. In the open air several tonnes are usually needed for an explosion but in a building a few tens of kilograms are sufficient. Enclosure increases explosion damage in two ways: it prevents dispersion of the leak and it prevents dissipation of the pressure generated by the explosion. It is also important to install gas detectors in places where men are not normally present as leaks may otherwise continue for a long time before they are detected.

In the present case we know that the leak was detected within 4 minutes of its start as, by chance, someone visited the area of the leak before it started and someone else saw the leak 4 minutes later. The procedure for shutting down the plant had been started but had not progressed very far when ignition occurred after a few minutes.

The building where the leak occurred was on two floors. The compressors were located on the upper floor and operators were normally present in this part of the building. The ground floor, where the leak occurred, housed various ancillary equipment and was normally visited about twice per hour. It was therefore a fortunate chance that the leak was detected so soon, and gas detectors were installed – on both floors – when the plant was rebuilt.

Gas detectors of an appropriate type should also be installed on plants handling toxic gases or vapours, particularly if experience shows that leaks are liable to occur.

Warn Three of the men killed by the explosion, and most of the injured men, were maintenance workers who were repairing a compressor. Although only a few minutes elapsed between the discovery of the leak and the ignition, there was ample time for them to leave the building. No-one told them to do so as no-one considered it possible that the leak would ignite. It was not normal practice to ask people to leave the building when a leak occurred. In fact, very small leaks were often ignored until it was convenient to repair them.

If a leak of flammable (or toxic) gas is detected, by people or by gas detectors, all people who are not required to deal with the leak should leave at once. There should be a recognised alarm signal, actuated automatically or by a responsible person, and a known assembly point, not too near the plant.

Isolate The fourth man killed was an operator who was attempting to shut down the plant. He at least had a reason for staying in the building when the leak was discovered. However, people should never be asked to enter or remain near a cloud of flammable gas in order to isolate a leak. Remotely operated emergency isolation valves should be provided. We cannot install them on the lines leading to and from all equipment which might leak but we can install them (a) when experience shows that leaks are liable to occur or (b) when, if leaks should occur, a large inventory will leak out. Examples of (a) are certain high pressure compressors, as in the present case, and very hot or cold pumps. Examples of (b) are the bottoms pumps on distillation columns[2-4].

In some cases, as well as remotely operated valves for isolating leaks, we also need remotely operated blow-down valves for getting rid of the inventory of hazardous material present in the section of plant that has been isolated. It may also be necessary to provide remote means of isolating power supplies such as pump and compressor drives. If a pump or compressor is fitted with remotely operated isolation valves, operation of these valves should automatically isolate the power supply.

If a remotely operated valve is fitted in the suction line of a pump or compressor, a non-return (check) valve may be installed in the delivery line instead of another remotely operated valve. This is acceptable provided that the valve is inspected regularly. Non-return valves have a bad name with many engineers but this may be because they are never looked at. No piece of equipment, especially one containing moving parts, can be expected to function for the lifetime of a plant without examination and repair if necessary. (On nuclear plants repair is often impossible and equipment has to be designed for lifetime operation but normal processing plants do not use such special equipment.)

Remotely operated isolation valves should not close too quickly or liquid hammer may overstress and even rupture the pipeline[5].

However many emergency isolation valves we install, a leak may occur on a piece of equipment which is not provided with them. Is it acceptable for someone to enter a cloud of flammable gas to isolate such a leak? I would not like to say 'Never'. I know of cases where a quick dash into a cloud has stopped a leak which might otherwise have continued for a long time. Anyone doing this should be protected by water spray and suitable clothing. However, we should try to avoid putting people into situations where they have this sort of decision to make, by generous provision of emergency valves. There were none on the plant we are discussing but they were installed when it was rebuilt. (Section 15.1, item

(6) describes another occasion when a man entered a vapour cloud to isolate a leak.)

Disperse 'The best building has no roof and no walls!' Whenever possible equipment handling flammable liquids and gases should be located in the open air so that small leaks are dispersed by natural ventilation. Not only is the dispersion better but much more gas or vapour, as already stated, has to be present before an explosion occurs. A roof or canopy over equipment is acceptable but walls should be avoided.

On the plant where the explosion occurred the chief engineer believed it would be difficult to maintain the compressors adequately if they were in the open and a closed building was installed on the rebuilt plant. However, in a new plant built a few years later, after he had retired, the compressors were protected only by a canopy. The company found that, like other companies, they could operate and maintain compressors successfully in the open air with no more than a roof over them. One company developed temporary walls and heaters, for use during maintenance[6].

The company owning the plant where the explosion occurred operated several other plants on the same site. Many of these had open compressor houses (though the compressors needed less attention than the ones we are discussing). When the explosion occurred a new compressor house had just been completed on one of these plants. It was near a workshop and so, in order to reduce the noise level in the workshop, the designers departed from their usual practice and built a closed compressor house. When they saw the report on the explosion they pulled down the walls before the compressor house was commissioned and dealt with the noise problem in other ways.

In recent years a number of enclosed compressor houses have again been built in order to reduce the noise level outside. There have been few explosions in recent years, the hazard has been forgotten and environmental regulations have become more stringent. Once again, we have solved one problem and created another[7]; once again we have forgotten the lessons of the past[8].

Perhaps modern compressors leak less often than old ones. However, a recent report shows how easily leaks can occur, often from pipework rather than the compressor itself. A compressor was partially enclosed, probably to aid winter operation and maintenance. Seven years earlier a spiral wound gasket had been replaced by a compressed asbestos fibre (caf) one. This was probably done as a temporary measure but once the caf gasket was installed it was replaced by the same type during subsequent maintenance. The gasket leaked and the leak exploded[9].

If, despite my advice, you build a closed compressor house, then forced ventilation is better than nothing but not nearly as good as the ventilation you get for free, even on a still day, in a building without walls.

Steam or water curtains can be used to confine and disperse leaks in the open air which otherwise might not disperse before they reach a source of ignition. Steam curtains have to be fixed but water curtains can be fixed or temporary. Temporary water curtains using fire service monitors have often been used to confine and disperse leaks and thus prevent them igniting.

Vent If equipment handling flammable gases or liquids has to be installed in a building then it is possible to minimise damage, if an explosion occurs, by building walls of such light construction that they blow off as soon as an explosion occurs, before the pressure has built up to a level at which it causes serious damage. Such walls do not, of course, protect people working in the building, as they will be killed or seriously injured by the fire, and they are therefore very much a second best choice. The walls have to be made of light plastic sheets fixed by special fastenings which easily break. In one case, when the designers specified such walls, the construction engineer was surprised that such weak fastenings had been requested. He did not know the reason for them. He therefore substituted stronger fastenings. When an explosion occurred the walls blew off at a higher pressure than the designers intended and damage was greater than it should have been.

Other lines of defence are fire-protection measures, such as insulation and water spray, and fire-fighting. They will not have any effect if an explosion occurs but may reduce the damage caused by a fire.

4.3   Reducing the probability of ignition

The source of ignition was never identified with certainty but two possible causes were found. Since either of them might have been the cause action had to be taken to try to eliminate both of them in the future.

Faulty electrical equipment The plant where the explosion occurred was classified as Zone (Division) 1 and the equipment was of flameproof design. Inspection of the equipment after the explosion, including sections of the plant not damaged by the explosion, showed that the standard of maintenance of the electrical equipment was poor. Gaps were too large, screws were missing, glasses were broken, unused entry holes were not plugged. A first glance, looking at equipment at eye level, suggested that nothing much was wrong. When equipment that could only be examined from a ladder was looked at, however, much of it was found to be neglected.

The action taken was similar to that taken to improve the standard of joint maintenance:

(1) It was found that many electricians did not understand the construction of flameproof equipment and did not realise the importance of correct assembly. New electricians were asked if they knew how to maintain the equipment. If they said 'No', a foreman gave them a half-

hour's demonstration. Training courses were set up, on other plants as well as on the plant where the explosion occurred.

(2) A system of regular inspection was set up. The inspectors examined a proportion of the equipment in use, selected at random, and some of the items inspected were dismantled. Similar inspections on other plants showed a sorry state of affairs, up to half the equipment being found faulty, though not all the faults would have made the equipment a source of ignition. Reader, before you criticise, have you examined your equipment?

(3) It was found that the special tools and bolts required for assembling flameproof equipment were not in stock. If an electrician lost a tool or dropped a bolt, what was he expected to do?

(4) On some plants, though not to a great extent on the plant where the explosion occurred, it was found that flameproof equipment was being used when Zone 2 equipment would be quite adequate. It requires much less maintenance than flameproof equipment and the cost of replacing flameproof equipment by Zone 2 equipment was soon recovered.

Static electricity Various bits of loose metal were found lying about the plant, mainly bits of pipe, valves and scaffold poles left over from maintenance work. If these were electrically isolated, as many of them were, and were exposed to a leak of steam or ethylene then a charge of static electricity might accumulate on them. When the charge reached a high enough level it could flash to earth, producing a spark of sufficient power to ignite an ethylene leak. Good housekeeping is essential, all bits of loose metal being removed. If they cannot be removed, for example, because they are required for a forthcoming turnaround, they should be grounded.

Other sources There was no evidence of smoking or other illegal sources of ignition. All employees recognised the need to eliminate smoking. There was no maintenance work going on in the ground floor area at the time of the explosion so a spark from a tool dropped by a maintenance worker on the concrete floor can be ruled out. The steam pipes in the basement were not nearly hot enough to ignite ethylene.

4.4   Incorrect knowledge

The explosion would not have occurred if the people who designed and operated the plant had realised that sources of ignition can never be completely eliminated even though we do what we can to remove all known sources. The fire triangle is misleading in an industrial context and instead we should say:

$$AIR + FUEL \rightarrow BANG.$$

We should therefore do everything possible to prevent and disperse leaks.

These statements were widely accepted on other plants on the same site. Explosions and fires had occurred on these plants in the past and lessons had been learnt from them. But these lessons had not been passed on to the plant where the explosion occurred or, more likely, they were passed on but nobody listened. The plant staff believed that their problems were different. If you handle ethylene as a gas at very high pressure it is perhaps not obvious that you can learn anything from an explosion on a plant handling propylene as liquid at a much lower pressure. Such an explosion had occurred in the same company, a few miles away, about 10 years earlier. The recommendations made, and acted upon, were very similar to those outlined above. But no one on the plant where the later explosion occurred took any notice. The plant was a monastery – a group of people isolating themselves by choice from the outside world – but fortunately the explosion blew down the monastery walls. Not only did the staff adopt many of the beliefs and practices current elsewhere, even, in time, building open compressor houses, but they developed a new attitude of mind, a much greater willingness to learn from the outside world.

The explosion illustrates the words of Artemus Ward (1834–1867), *'It ain't so much the things we don't know that get us in trouble. It's the things we know that ain't so'.*

Monastic attitudes, however, continued elsewhere in the company. Before the explosion some low pressure ethylene pipework had been constructed of cast iron. This suffered much greater damage than steel pipework and it was agreed that cast iron would not be used in future. Only steel was used in the rebuilt plant. A couple of years later construction of a plant which used propylene was sanctioned. The design team insisted that rules agreed for ethylene did not necessarily apply to them and the arguments against cast iron had to be gone through all over again.

Similarly, although the recommendations made after the explosion, particularly the need for open compressor houses, were circulated throughout the company, little attention was paid to them on plants that handled hydrogen. Hydrogen, it was argued, is much lighter than any other gas and disperses easily. A few years later the roof and walls were blown off a hydrogen compressor house (Figure 4.2).

Individual parts of the company were allowed considerable autonomy in technical matters. Many years before it had been formed by an amalgamation of independent companies who still cherished their freedom of action and it was felt that attempts to impose uniform standards and practices would lead to resentment. The explosion did not produce a re-examination of this philosophy though perhaps it should have done. It probably never occurred to the senior managers of the company that their organisational structure had any bearing on the explosion. They knew it was due to a badly made joint and so no expense was spared to improve the standard of joint making. The management philosophy was not changed until many years later when recession caused different parts of the company to be merged.

Figure 4.2 The result of an explosion in a hydrogen compressor house

However, a word of caution. Because the senior managers of the company might have prevented an accident by changing their organisation and philosophy, we should not use this as an excuse for doing less than possible at other levels. The chain of causation could have been broken at any level from senior manager to craftsman. The accident might not have occurred if the organisation had been different. Equally it would not have occurred if the joint had been properly made.

Other explosions in closed buildings are described in references 10–12.

4.5   Supercritical gases may behave like liquids

Although ethylene at high pressure and atmospheric temperature is a gas, being above its critical temperature, it behaves in some ways like a flashing liquid, that is, a liquid under pressure above its normal boiling point. Thus its density is close to that of a liquid and a small leak produces a large gas cloud. The precautions which should be taken when handling it are similar to those that should be taken when handling liquefied flammable gases such as propylene. There was more to be learned from the propylene explosion ten years earlier than the staff realised.

4.6   Postscript

After I had completed this chapter I came across a memorandum on 'Safety Precautions on Plants Handling Ethylene', issued thirty years

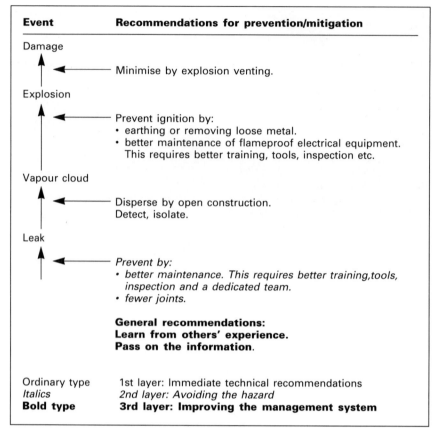

<figure>

Event	Recommendations for prevention/mitigation
Damage	
↑ ←	Minimise by explosion venting.
Explosion	
↑ ←	Prevent ignition by: • earthing or removing loose metal. • better maintenance of flameproof electrical equipment. This requires better training, tools, inspection etc.
Vapour cloud	
↑ ←	Disperse by open construction. Detect, isolate.
Leak	
↑ ←	*Prevent by:* • *better maintenance. This requires better training,tools, inspection and a dedicated team.* • *fewer joints.* **General recommendations:** **Learn from others' experience.** **Pass on the information.**
Ordinary type *Italics* **Bold type**	1st layer: Immediate technical recommendations *2nd layer: Avoiding the hazard* **3rd layer: Improving the management system**

</figure>

Figure 4.3 Summary of Chapter 4 – A gas leak and explosion – The hazards of insularity

before the explosion, by the predecessor company, when they started to use ethylene. It said, 'In all the considerations that have been given in the past to this question, it has always been felt that excellent ventilation is a *sine qua non*. Not only does it reduce the danger of explosive concentrations of gas occurring, but it also protects the operators of the plant from the objectionable smell and soporific effects of ethylene'.

During the thirty years this sound advice was forgotten and ignored. Chapters 5 and 13 and reference 8 describe other accidents which occurred because knowledge was lost.

References

1 Urben, P. G., *Journal of Loss Prevention in the Process Industries*, Vol. 2, No. 1, Jan. 1989, p. 55.
2 Kletz, T. A., *Chemical Engineering Progress*, Vol. 71, No. 9, Sep. 1975, p. 63.

3 Kletz, T. A., *What Went Wrong? Case Histories of Process Plant Disasters*, 2nd edition, Gulf Publishing Co., Houston, Texas, 1988, Section 7.2.

4 Lees, F. P., *Loss Prevention in the Process Industries*,Vol. 1, Butterworths, London, 1980, Section 12.6.2.

5 Lees, F. P., *Loss Prevention in the Process Industries*, Vol. 1, Butterworths, London, 1980, Section 12.6.4.

6 Morris, D. H. A., *Loss Prevention in the Process Industries*, edited by C. H. Buschmann, Elsevier, Amsterdam, 1974, p. 369.

7 Kletz, T. A., *Process Safety Progress*, Vol. 12, No. 3, July 1993, p. 47 .

8 Kletz, T. A., *Lessons from Disaster - How Organisations have No Memory and Accidents Recur*, Institution of Chemical Engineers, Rugby, UK, 1993.

9 MacDiarmid, J. A. and North, G. J. T., *Plant/Operations Progress*, Vol. 8, No. 2, April 1989, p. 96.

10 Howard, W. B., *Loss Prevention*, Vol. 6, 1972, p. 68.

11 *The Fire and Explosions at Permaflex Ltd, Trubshaw Cross, Longport, Stoke on Trent, 11 February 1980*, Health and Safety Executive, London, 1981.

12 *Vigilance*, Vol. 4, No. 7, Summer 1987, p. 70.

Chapter 5

A liquid leak and fire – The hazards of amateurism

It is beginning to be hinted that we are a nation of amateurs.

Lord Rosebery (1847–1929),
foreign secretary and prime minister, speaking in 1900

Two men were killed and several seriously injured when a leak of 4 tonnes of hot flammable hydrocarbon ignited. There was no explosion, only a fire which was soon extinguished. Material damage was relatively light – about £1.5 million at 1994 values – but the consequential loss (business interruption) was thirty times this figure.

The hydrocarbon processing plant consisted of two parallel units which shared some equipment in common (Figure 5.1). One of the units was shut down for maintenance while the other continued on line. The unit under repair should have been isolated by slip-plates (spades) but unfortunately one of the interconnecting lines was overlooked. As a result the leak occurred.

We shall therefore look at:

- The action needed to prevent similar leaks in the future.
- Ways of reducing the probability of ignition.
- Ways of minimising the effects of the fire.
- The influence of the management experience and philosophy.

First, however, let us look at Figure 5.1 in more detail. The reaction sections of the two units were quite separate. Reaction product entered a common product receiver and was then further processed in a common unit. The product receiver could be by-passed by the lines shown dotted in Figure 5.1. These lines had been installed to make it easier to empty the reactors when they were being shut down and were used only occasionally. They were therefore overlooked when slip-plates were inserted to isolate the operating unit from the one that was to be maintained.

Valve B, in the by-pass line from the shutdown unit (No 2), had been removed for overhaul. A few hours later valve A was operated. Hot

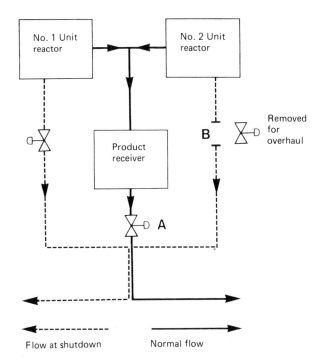

No. 1 Unit reactor

No. 2 Unit reactor

Product receiver

B

Removed for overhaul

A

Figure 5.1 Simplified flow diagram showing a combination of series and parallel equipment

Flow at shutdown   Normal flow

hydrocarbon (boiling point about 85°C, temperature about 150°C and gauge pressure about 7 bar [100 psi]) travelled in the wrong direction along No 2 unit by-pass line and came out of the open end. Someone sounded the evacuation alarm and the numerous maintenance workers who were on the plant at the time, working on No 2 unit, started to leave. Unfortunately, for some of them the direct route to the main gate and assembly point lay close to the leak. As they were passing it, ninety seconds after the leak started, the hydrocarbon ignited. However, most of the eighty maintenance workers who were on the plant at the time escaped without injury.

Valve A had been operated remotely from the control room. There were no windows and the operator could not see the plant. He heard the evacuation alarm sound. He did not know the reason but as he had just operated valve A, he thought this might be connected in some way with the alarm and he shut it again. His prompt action prevented a much more serious fire.

Because the hydrocarbon was under pressure above its normal boiling point a very large proportion of the leaking material turned to vapour and spray. The fire was therefore much more extensive than that which would have followed the spillage of 4 tonnes of cold hydrocarbon. The initial flash fire covered a substantial area of the plant and this was followed by a smaller fire close to the point of leakage.

5.1   Prevention of the leak

The policy on the plant where the fire occurred was that equipment under repair should be isolated by slip-plates (or disconnection of a section of pipework) unless the repair job was a quick one. If a whole unit or section of a unit was shut down for maintenance then that unit or section could be isolated as a whole and it was not necessary to isolate each individual item of equipment. This was a sound policy and it was intended to follow it but unfortunately a little-used interconnecting line between the two units was overlooked. Although a permit-to-work was issued authorising the maintenance organisation to separate the two units it did not list each joint to be slip-plated or disconnected but merely said, 'Insert slip-plates or disconnect joints to separate No. 1 unit from No. 2 unit'. The plant was not, of course, as simple as implied by Figure 5.1; it was a 'spaghetti bowl'.

Chapter 3 described an accident which occurred because a fitter was asked to remove 'all' the slip-plates from a vessel and left one in position. In the incident described in this chapter the foreman who was asked to decide where the slip-plates should be placed overlooked one of the connecting lines.

After the fire the following recommendations were made to prevent a similar incident in the future.

(1) A schedule of isolations should be prepared well in advance of a shutdown, using the line diagrams, and checked on the ground by tracing all interconnecting lines. Particular care should be taken when tracing lines that are insulated together. Line diagrams should be kept up-to-date.

(2) The joints to be slip-plated (or disconnected) should be marked with a numbered tag and listed individually on a permit-to-work. It is not sufficient to write on the permit, 'Fit slip-plates to separate No 1 unit from No 2 unit' or 'Slip-plate all inlet and exit lines on No 2 unit'.

(3) If a plant contains two (or more) closely interconnected units then, whenever possible, one should not be maintained while the other is on line. If it is essential to do so then the two units should be widely separated physically and the isolations necessary should be agreed during design, not left for a foreman to sort out a few days before a shutdown.

(4) Hazard and operability studies (hazops)[1-3] should be carried out on all new designs and they should include a check that all equipment or sections of plant that may be maintained while the rest of the plant is on line can be simply and safely isolated.

These recommendations apply to all plants handling hazardous materials and not just to the plant where the fire occurred.

5.2 The source of ignition

The source of ignition was an unusual one and attracted widespread interest at the time. The company concerned issued a press release and reports appeared in several journals[4]. As a result little interest was shown, outside the company, in the reasons why the leak had occurred.

A diesel-engined vehicle was being used by the maintenance team who were working on the unit that was shut down. When the leak occurred hydrocarbon vapour was sucked into the engine through the air inlet and the engine started to race. The driver tried to stop the engine in the normal way, by isolating the fuel supply, but as the engine was getting its fuel through the air inlet it continued to race. Valve bounce occurred and the vapour cloud was ignited by flash-back from the cylinders.

The company concerned did not allow petrol (gasoline) engines into areas where flammable gases or liquids were handled but diesel engines were considered safe as they produced no sparks. The fire showed that they are as dangerous as petrol engines and should be treated with the same respect. In fact, several other leaks had been ignited by diesel engines in the same way, and in other ways, but they had received little publicity.

Since the time of the fire several proprietary devices have been devised which make it possible to shut down a diesel engine which is sucking in fuel through the air inlet. A valve in the air inlet closes and in addition a fire extinguishing agent such as carbon dioxide or Halon is injected into the air inlet. The devices can be operated by hand or they can be operated automatically by a flammable gas detector. (The use of Halon is being phased out as it damages the ozone layer.)

Diesel engines can ignite leaks in several other ways and if we wish to prevent them doing so the exhaust temperature should be kept below the auto-ignition temperature of the gas or vapour, a spark arrester and flame arrester should be fitted to the exhaust and the decompression control should be disconnected. The electrical equipment should be suitable for the area in which the equipment will operate, that is, Zone (Division) 1 or 2.

The following recommendations, of general applicability, were made after the fire[5]. They do not go as far as the recommendations made for the use of diesel engines in mines[6].

(1) Diesel engines which operate for more than 1000 hours per year in Zone (Division) 1 or 2 areas should be fully protected with a device that enables them to be stopped if fuel is being sucked in through the air inlet, spark and flame arresters on the exhaust and suitable electrical equipment. Compressed air or spring starters should be used instead of electric starters. The exhaust temperature should be kept below the auto-ignition temperature of the materials which might leak and the decompression control should be disconnected. Fixed diesel

engines should whenever possible be located in safe areas or, if this is not possible, their air supplies should be drawn from safe areas and their exhausts should discharge to safe areas, through long lines if necessary.

(2) Vehicles which are used occasionally, for example, during maintenance operations, can be protected to a lower standard if they are never left unattended. They should, however, be fitted with a device which enables them to be shut down if fuel is being sucked in through the air inlet, the exhaust temperature should be below the auto-ignition temperature of the materials which might leak and the decompression control should be disconnected. If there are no flammable gas detector alarms in the area then portable ones should be installed.

(3) Vehicles which are just passing through, for example, delivering or collecting materials, need not be protected at all but should not enter without permission from the process foreman. This permission should not be given unless conditions are steady and there are no leaks.

The effect of these recommendations was to discourage the use of cranes, welding sets, etc., in Zone (Division) 1 and 2 areas while plants are on line, while falling short of stopping it completely. When cranes were used, the use of hydraulic cranes in place of diesel electric cranes was encouraged.

5.3   Minimising the effects of the fire

The plant was a 'No Smoking' area and was therefore surrounded by a fence to control access. Only one gate was in regular use. As already stated the direct route to this gate and the prearranged assembly point, for some of the eighty maintenance workers, lay close to the leak. When the evacuation alarm sounded these men started to leave and were passing close to the leak when it ignited. Other men left in other directions by climbing the fence.

After the fire the rules were changed. Additional gates, which could be opened only from the inside, were installed and employees were told to leave by the nearest gate and make their way round the plant, if necessary, to the assembly point.

The fire drew attention to the fact that eighty men, in addition to the normal operating team, were working on a plant that was in partial operation. This is bad practice, as already stated, and plants should not be designed with closely-interconnected parallel units if the units are to be overhauled at different times.

The congested design of the plant made rapid evacuation difficult and increased the damage caused by the fire. In addition the ground was sloped so that spillages flowed towards the centre of the plant instead of away from it (as in the accident described in Chapter 3). Much of the drainage flowed along open channels instead of underground drains. Both these features encouraged the spread of fire and considerable sums were spent afterwards in trying to alleviate them. It would have cost little more at the design stage if the ground had been sloped so that spillages ran away from the main items of equipment and drains were underground.

As already stated, the control room operators could not see the plant but they heard the evacuation alarm, one of them assumed that his last action might have been responsible, and reversed it. This prompt action prevented a more serious fire. The probability that an operator will take the correct action in such a situation may be increased if he can see the plant through the control room windows. Windows are a weak spot if a control room is being strengthened to resist blast but nevertheless many people believe that they are worth having in order to give the operators a view of the plant. In theory, all the information they need to control the plant can be obtained from instruments or closed circuit television. In practice, many operators feel that a view of the plant is helpful when things are going wrong. Windows need not be very big (up to 1 m^2) and should be made of shatter-resistant glass or glass protected with plastic film.

5.4 The management experience and philosophy

If the explosion described in Chapter 4 was due to a blinkered attitude – deliberate isolation from the rest of the organisation – the fire just described was due to amateurism. The company involved had been engaged in batch chemicals production for many decades and were acknowledged experts. Sales of one product, which had started as a batch product, had grown so much that construction of a large continuous plant was justified, in fact several such plants had been built. The company had little experience of continuous production so they engaged a design contractor and left it to him. Unfortunately the contractor they chose produced a good process design but a poor engineering design. The plant was extremely congested, thus aggravating the effects of the fire, and the combination of single stream and parallel operation, with the two streams closely interconnected, both physically and in the line-diagram sense, made maintenance difficult and increased the likelihood of error. The drainage arrangements were almost scandalous.

The company was part of a larger group, and many of the other companies in the group were involved in the design and operation of continuous plants. They had learnt by bitter experience that they should not engage a contractor, however competent, and leave the design to him. They monitored the design closely as it developed. They had found that employment of a contractor did not reduce the amount of graduate engineer effort required, though it did, of course, vastly reduce the draughtsman effort required. At no time did the staff of the batch company seek help or advice from the other companies in the group. It

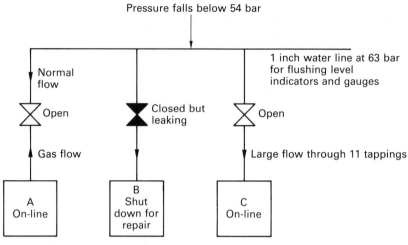

Gasifiers – operating pressure 54 bar

Figure 5.2 Another accident due to a missing slip-plate. When the pressure in the water line fell, gas from gasifier A entered the water line, passed through the leaking valve into gasifier B and exploded

probably did not even occur to them that they might usefully seek advice. Like the unfortunate engineer at Flixborough who designed the pipe which failed (Chapter 8) they did not know what they did not know. The managers of the holding company allowed its component parts considerable autonomy in technical matters and never commented on choice of contractor or said who should be consulted. However, about two years after the fire there was a major re-organisation and responsibility for the plant involved in the fire, and other similar plants, was transferred to another part of the group.

As with the accident described in Chapter 4, it probably never occurred to the senior managers of the company that their actions, or inactions, had contributed towards it, that if they had taken advice from those more experienced in handling large continuous plants it might never have occurred. But again the chain could have been broken at any point. The accident would not have occurred if the senior managers had been less insular; equally, it would not have occurred if the foreman responsible for seeing that the slip-plates were fitted had inspected the plant more thoroughly. However, the foreman had more excuse; he was working under pressure. The senior managers had time to reflect on their decisions.

5.5   Consequential loss

As stated at the beginning of this chapter, the fire caused comparatively little material damage – mainly instruments and cables – but replacing

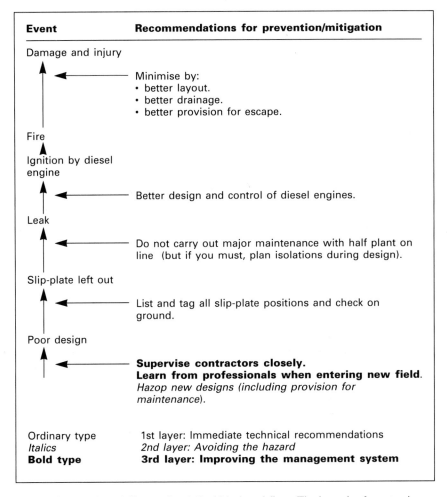

Event	Recommendations for prevention/mitigation

Damage and injury
— Minimise by:
• better layout.
• better drainage.
• better provision for escape.

Fire

Ignition by diesel engine
— Better design and control of diesel engines.

Leak
— Do not carry out major maintenance with half plant on line (but if you must, plan isolations during design).

Slip-plate left out
— List and tag all slip-plate positions and check on ground.

Poor design
— **Supervise contractors closely.**
Learn from professionals when entering new field.
Hazop new designs (including provision for maintenance).

Ordinary type — 1st layer: Immediate technical recommendations
Italics — *2nd layer: Avoiding the hazard*
Bold type — **3rd layer: Improving the management system**

Figure 5.3 Summary of Chapter 5 – A liquid leak and fire – The hazards of amateurism

them took a long time and the consequential loss of profit was thirty times the material damage. The company had taken out consequential loss (business interruption) insurance only a short time before the fire and were glad that they had done so. The insurance companies were surprised at the extent of the business interruption caused by the combustion of four tonnes of hydrocarbon and their premiums for this type of insurance rose steeply.

5.6   Another incident

Another incident involving shared equipment occurred on an ammonia plant on which there were three parallel gasifiers, A, B and C[7] (Figure

5.2). They were distinct units but a common 1 inch (25 mm) line supplied water at a gauge pressure of 63 bar (910 psi) for flushing out level indicators and level gauges. The plant operated at a gauge pressure of 54 bar (780 psi).

Gasifiers A and C were on line and B was shut down for repair. A large amount of flushing water was used on C and the pressure in the water line near A and B fell below 54 bar. Gas from A entered the water line, entered B through a leaking valve and came out inside B where it was ignited by a welding spark. Fortunately no one was injured.

The connection from the water line to B gasifier had not been slip-plated as no one thought it possible for gas to enter by this route. Afterwards separate water lines were run to the three gasifiers. Chapters 2 and 11 describe other incidents which started when backflow occurred into a service line.

References

1   Kletz, T. A., *Hazop and Hazan – Identifying and Assessing Process Industry Hazards*, 3rd edition, Institution of Chemical Engineers, Rugby, UK, 1992.
2   Lees, F. P., *Loss Prevention in the Process Industries*, Vol. 1, Butterworths, 1980, Chapter 8.
3   Knowlton, R. E., *A Manual of Hazard and Operability Studies*, Chemetics International, Vancouver, Canada, 1993.
4   For example, *Chemical Age*, 12 December 1969, p. 40 and 9 January 1970, p. 11.
5   Oil Companies Materials Association, *Recommendations for the Protection of Diesel Engines Operating in Hazardous Areas*, Wiley, Chichester, UK, 1977.
6   British Standard 6680:1985, *Flameproof Equipment for Diesel Engines for use in Coal Mines and other Mines Susceptible to Firedamp*.
7   Mall, R. D., *Plant/Operations Progress*, Vol. 7, No. 2, April 1988, p. 137.

Chapter 6

A tank explosion – The hazards of optional extras

We have become amphibious in time. We are born into and spend our childhood in one world; the years of our maturity in another. This is the result of the accelerating rate of change.

B. W. Aldiss.

An explosion followed by a fire occurred in a 1000 m³ fixed roof storage tank which was one-third full of a volatile hydrocarbon (Figure 6.1). The roof was blown off the tank but remained attached at one point. Application of the theoretical quantity of foam from hoses failed to extinguish the fire, either because the foam dried out before it reached the

Figure 6.1 The inside of the tank after the fire

49

surface of the burning liquid or because it was swept upwards by the rising plume of combustion products. However, a monitor delivering 23 m³/min (5000 gall/min) of foam extinguished the fire in ten minutes. Nobody was injured. The hydrocarbon (flash point 45°C) was unconverted raw material recovered from a product by distillation and was contaminated with 0.5% of a more volatile liquid which lowered the flash point to near ambient. The contaminant had a high conductivity. The tank was being used on balance, liquid being added to the tank and withdrawn for recycling at about the same rate (2 m³/h).

For a fire we need fuel, air (or oxygen) and a source of ignition and we shall consider these separately before looking at the weaknesses in the management systems. It is also useful to ask, when considering a fire or explosion, or indeed any accident, why it happened when it did and not at some other time.

6.1   The fuel

Despite the contamination of the liquid in the tank with 0.5% of a more volatile material its flash point was still above the ambient temperature. However, as a result of an upset on the distillation column the amount of contaminant in the incoming stream rose to 9% and the flash point fell. If this incoming liquid had been added to the base of the tank, the usual practice when adding liquid to a tank, the change in the composition of the liquid would have had little effect on the composition of the vapour. But the tank was fitted with a swing arm so that liquid could be withdrawn from any level. This swing arm was stuck in the fully raised position and could not, therefore, be used for withdrawing liquid from the tank which was one-third full. The swing arm was therefore being used for delivery and the normal delivery line into the base of the tank was being used for

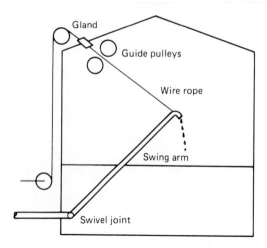

Figure 6.2 The tank was being splash-filled through a swing arm which was stuck in the raised position. (Drawing not to scale)

suction (Figure 6.2). As a result the composition of the vapour space reflected the composition of the incoming, splashing liquid rather than the composition of the bulk of the liquid in the tank and the vapour composition changed as soon as the composition of the incoming liquid changed. This probably explains why the explosion occurred soon after the upset on the distillation column.

The unit manager knew that it was bad practice to splash fill a tank but he felt he had no other option and he did not consider the risk serious as he thought the tank was blanketed with nitrogen. He had arranged for a spare branch near the base of the tank to be connected up so that it could be used for delivery and a plumber arrived to make the modification while the tank was on fire!

6.2   The air

It was company policy to blanket with nitrogen all fixed roof storage tanks containing hydrocarbons with flash points below or close to ambient temperature and the tank which exploded should have been blanketed. The system used was old-fashioned but quite effective if used correctly. All the fixed roof storage tanks were connected to a gasholder. When liquid was added to a tank nitrogen was pushed into the gasholder; when liquid was withdrawn from a tank, nitrogen was withdrawn from the gasholder. This system conserved nitrogen but today the value of the saving would not justify the cost of installing a gasholder. Customs and Excise regulations required the tanks to be manually dipped every month

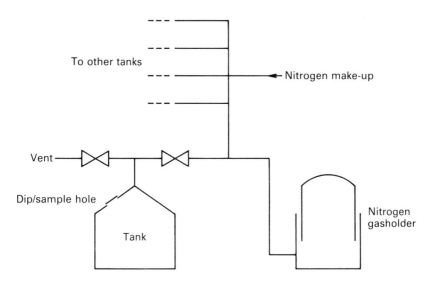

Figure 6.3 The tank was connected to the nitrogen gasholder but could be isolated and vented while it was dipped

and to do this they had to be disconnected from the gasholder and vented to atmosphere for a short time, otherwise nitrogen would be wasted and would also be discharged from the dip hole close to the dipper (Figure 6.3).

After the explosion it was found that the tank was disconnected from the gasholder and open to the atmosphere. The unit manager had personally inspected the tank seven months before the explosion, when it had been first used on its current duty, and at the time the valves on the tank had been set correctly. It is believed that during the intervening months the operator who dipped the tank had failed to reconnect it to the nitrogen system and that thereafter other operators had left it as they found it. No operator would admit that he found or left the valves in the wrong position. The manager and foremen had not looked at them during the seven-months period.

It was perhaps as well that the tank was disconnected from the nitrogen blanketing system as the nitrogen in the gasholder was found to contain 15% oxygen. If the tank had been connected to the gasholder and other tanks then the explosion might have spread through the connecting lines to every tank on the site. For once, two wrongs did make a right!

How did the air get into the nitrogen system? It is believed that it leaked in through tank sample and dip holes which had been left open and through an open-end which was discovered on one of the storage tanks. In addition it is possible that the gasholder may have been allowed to get down to its minimum position in which case air would have been sucked into the tanks through their pressure/vacuum valves (conservation vents).

After the explosion a special tank inspector was appointed to check the position of the nitrogen valves on every storage tank every day. He carried a portable oxygen analyser and also checked the composition of the atmosphere in each tank. There were many tanks on the site and it was considered that the foremen would not have time to personally inspect them regularly. In addition, it would have been a good idea to have installed a permanent oxygen analyser alarm on the gasholder.

6.3   The source of ignition

Several sources of ignition were considered and eliminated. At first static electricity was considered to be the obvious source but the filling rate was so low and the conductivity of the liquid so high that sufficient charge could not accumulate. A discharge from a mist of oil droplets was considered but an experiment (with water) showed that liquid falling the distance that it actually fell at the actual flow rate (2 m^3/h) did not produce any significant amount of mist. The walls of the tank were examined for traces of pyrophoric or catalytically active material and the liquid was examined for traces of peroxides but nothing was found. External sources of ignition could be ruled out as there was no welding or other hot work in progress anywhere near the tank at the time, the weather was fine and the tank was located in a 'No Smoking' area.

A possible source of ignition that could not be ruled out is based on an observation by the Safety in Mines Research Establishment[1] that when a wire rope is subjected to friction small 'hairs' rubbed off the wire become incandescent. These hairs will not ignite methane but it is possible that they could ignite the hydrocarbon or the contaminant which both had a much lower auto-ignition temperature than methane. The swing-arm, which, as already stated, was stuck in the upright position, was supported by a wire rope which passed over a pulley inside the tank. The pulley was seized and the wire rope had been tightened to try to reduce the vibration of the swing-arm. It is therefore possible that the heat or frictional 'hairs' produced by the vibration of the rope against the pulley provided the source of ignition.

The incident confirms the statement made in Chapter 4 that possible sources of ignition are so numerous that we can never be certain that we have eliminated them completely, even though we try to remove all known sources. Flammable mixtures should never be tolerated except in a few special cases when the risk of ignition is accepted. One such case is the vapour spaces of fixed roof storage tanks containing liquids of high conductivity such as alcohols and ketones as there is little or no chance of ignition by static electricity, if the tanks are earthed[2].

The company concerned required fixed roof tanks containing volatile hydrocarbons to be blanketed with nitrogen, as many explosions in such tanks have been reported in the literature[3], but did not require blanketing on tanks containing high-conductivity liquids as very few explosions have been reported in such tanks. Although the tank which exploded contained hydrocarbon, the contaminant gave it a high conductivity. Should the policy therefore be revised? The company decided to continue with its policy of not blanketing tanks containing conducting liquids but to make it clear that splash filling and any form of mechanical movement or vibration could not be accepted. They did not have many tanks connected by a common vent system but it was clear that such a system provided an ideal means of spreading an explosion from one tank to another and that any tanks connected to such a system should be blanketed.

6.4   An unlikely coincidence?

As in the accident described in Chapter 2, at first sight the explosion seems to have been the result of an unlikely coincidence, the loss of nitrogen blanketing, the rise in concentration of the volatile component and the source of ignition all occurred at the same time. Who could have foreseen that? In fact, the scene was set by three ongoing unrevealed or latent faults and when a fourth occurred, the explosion was inevitable. The three ongoing faults were:

- The nitrogen blanketing was out of use and probably had been for months.
- Splash filling had been in use for months.
- The source of ignition had probably been present for some time.

When the concentration of the volatile component in the inlet stream rose, as it was likely to do sooner or later, the splash filling ensured that the concentration in the vapour space also rose and an explosion was inevitable.

6.5   Limitation of damage

Fortunately the tank was provided with a weak seam roof. The roof/wall weld failed first, as it was supposed to do, and there was no spillage. The fire burned harmlessly in a confined space and caused little damage. If the tank had failed elsewhere burning liquid would have filled the bund which was shared with other tanks. The incident therefore shows the value of weak seam roofs.

A flare stack was located near the tank. Two guy ropes supporting this flare stack passed over the tank. The lower rope, which was close to the tank, was broken by the roof when it blew off. Fortunately the upper rope, though exposed to the heat, did not fail. If it had, the stack might have collapsed. Storage areas and operating plant should be better segregated.

6.6   Training and inspections

The explosion would not have occurred if the nitrogen blanketing system had been kept in working order. For any protective system to be kept in working order, those who operate it must be convinced of its value and regular checks must be made to make sure that it is in working order. These two actions go together. If managers do not make regular checks, then everyone assumes that it cannot be very important.

It is clear that operators, foremen and perhaps junior managers did not regard nitrogen blanketing as a major safety precaution. It was looked upon as an 'optional extra', to use a phrase from motor car catalogues: a luxury extra to adopt if you can but something that can be dropped if you are busy or nitrogen is scarce. In contrast, everybody regarded the prohibition of smoking as an important safety measure and there was no difficulty enforcing it. Incidents of smoking were almost unknown and the few that had occurred were confined to contractors, not regular employees.

The nitrogen blanketing system had been installed many years before. It had fallen into disuse but five years before the explosion, a survey of tank explosions had been made and a decision taken to bring the nitrogen blanketing back into use. The necessary instructions were issued but no attempt was made to explain to employees, at any level, why the change had been made. We do not live in a society in which management by edict is possible. People want to know why they are asked to do things, particularly when they are asked to change their habits. The management had failed to provide this explanation. A written explanation would not have been sufficient. Any such change in policy should be discussed with those who will have to carry it out and their comments obtained.

But explanations are not enough. They must be followed by regular audits or inspections to make sure that the new practices are being followed. These also the management had failed to provide. After the explosion, as already stated, they appointed a full-time tank inspector, as there were so many tanks to be checked, and managers themselves made occasional checks.

This incident has been discussed on many occasions with groups of managers and design engineers, using the method described in Part 4 of the Introduction. Sometimes they comment on the rather old-fashioned nature of the nitrogen blanketing system and suggest a more modern once-through system. However, the old-fashioned system was quite safe if used properly. If people did not use the equipment provided, would they have used better equipment? Nevertheless, a once-through system is better as each tank is then isolated and if anything goes wrong with the blanketing on one tank it does not spread to the other tanks. Reference 4 describes an explosion in a pair of tanks which shared a common nitrogen system.

At Bhopal (Chapter 10) much of the safety equipment was out-of-use.

As already described, the tank had to be disconnected from the nitrogen system every month so that it could be dipped. This was tedious and so there was a temptation for operators to leave the tank disconnected ready for the next month. If the safe method of operation is more troublesome than the unsafe method then many people will follow the unsafe one. Systems should be designed so that safe operation is not difficult. If a dip-pipe had been installed in the tank it would have been possible to dip it without disconnecting it from the nitrogen system.

6.7   Other incidents

Explosions can spread from one tank to another through common vent systems as well as common nitrogen systems. In 1991 many tanks in a chemical storage depot at Coode Island, Melbourne, Australia were destroyed by fire. Whatever the source of ignition (sabotage has been suggested[5]), it seems that the common vent system allowed the fire to spread from one tank to another, either directly through the vent lines or as a result of their exposure to fire.

The storage installation was originally constructed without a common vent system but one was installed later, and connected to a vapour recovery unit, to prevent discharge of vapours to atmosphere. A decision not to blanket with nitrogen was acceptable when each tank stood alone but not when they were connected together[6]. Once again we see a lesson learned once, learned again several decades later, though on the second occasion the 'tuition fees' were higher[7]. Once again we see the unpleasant and unforeseen side-effects of a change made with the commendable intention of improving the environment[8].

6.8   Avoiding the problem

Was it necessary to store the recovered hydrocarbon or to store so much? In the original design it was recycled as it was produced and there was no intermediate storage. The unit manager decided that buffer storage would make operation easier and as a spare storage tank was available he made use of it. He probably gave no thought to the increase in hazard or estimated the minimum quantity he could manage with. Storage of flammable liquids in tanks was an accepted hazard; he was making little increase in the total quantity on site and if he had not used the tank for his purpose somebody else would have put something else in it. Nevertheless large stocks are made up of smaller stocks and the need to keep storage to the minimum (for economic reasons as well as safety ones) should always be emphasised. 'What you don't have, can't explode.'

The accident at Bhopal in 1984, which killed 2000 people (Chapter 10) was also due to unnecessary storage of a hazardous intermediate.

6.9   Seeing the signal above the noise

This explosion occurred in the 1960s. Nevertheless all the lessons that can be learned from it, described above, are still relevant. Many accidents still occur because protective equipment is neglected or hazards are not explained to those who have to control them. These lessons were learnt in the company concerned. In addition the incident had a wider effect. The company had a good safety record and attached great importance to safety but it was considered a non-technical subject. The safety officers were mainly non-technical and concerned themselves with mechanical hazards, protective clothing and similar matters. The explosion was the first of a series of accidents which were to occur in the next few years and which gradually led the management to realise that there was a need for a technical input to safety, for a new sort of safety adviser who understood the technology and could advise on the action needed to prevent the sort of accidents which were now occurring, accidents similar to those described in Chapters 4, 5 and 7 of this book.

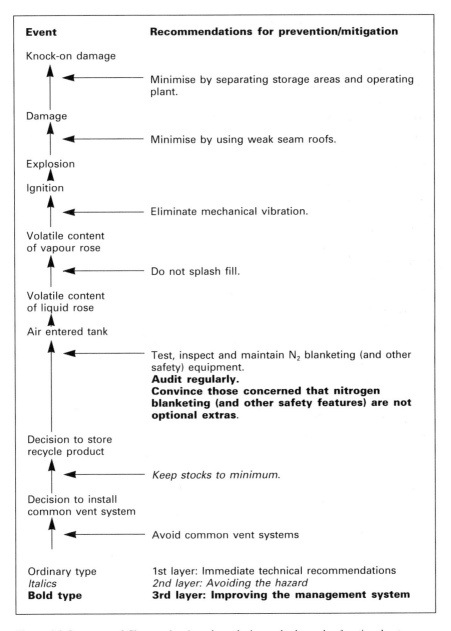

Event   **Recommendations for prevention/mitigation**

Knock-on damage

Minimise by separating storage areas and operating plant.

Damage

Minimise by using weak seam roofs.

Explosion

Ignition

Eliminate mechanical vibration.

Volatile content of vapour rose

Do not splash fill.

Volatile content of liquid rose

Air entered tank

Test, inspect and maintain N$_2$ blanketing (and other safety) equipment.
Audit regularly.
Convince those concerned that nitrogen blanketing (and other safety features) are not optional extras.

Decision to store recycle product

Keep stocks to minimum.

Decision to install common vent system

Avoid common vent systems

Ordinary type	1st layer: Immediate technical recommendations
Italics	*2nd layer: Avoiding the hazard*
Bold type	**3rd layer: Improving the management system**

Figure 6.4 Summary of Chapter 6 – A tank explosion – the hazards of optional extras

What had changed? This is a matter of conjecture but the following factors were probably important: First, during the 1950s a new generation of plants had been built which were larger than those built before and operated at higher temperatures and pressures; leaks, if they occurred, were therefore more serious. Second, plants were more dependent on instrumented protective systems but managers were slow to realise the amount of testing and maintenance that was required. Most of the operators had grown up on an earlier generation of plants and were slow to adapt to new ways. Managers did not realise the amount of retraining that was required.

Whatever the reasons for the changes, there is no doubt that it took several years and several serious fires and explosions before it was generally realised in the company that there had been a step change. Each fire and explosion was followed by a series of recommendations designed to remove the immediate technical causes, and to improve such matters as testing and inspection, but it was several years before it was realised that an entirely new approach, the loss prevention rather than the traditional safety approach, was needed. It is always difficult to see the signal above the noise when we do not know what signal to expect and we hope there is no signal at all.

References

1 Powell, F. and Gregory, R., *Technical Report No 41*, Safety in Mines Research Establishment, Sheffield, UK, 1967.
2 Kletz, T. A., 'Hazard analysis - A quantitative approach to safety', *Major Loss Prevention in the Process Industries*, Symposium Series No 34, Institution of Chemical Engineers, Rugby, UK, 1971, p. 75.
3 Klinkenberg, A. and van der Minne, J. L., *Electrostatics in the Petroleum Industry*, Elsevier, Amsterdam, 1958.
4 Kletz, T. A., *What Went Wrong? Case Histories of Process Plant Disasters*, 2nd edition, Gulf Publishing Co., Houston, Texas, 1988, Section 5.4.2.
5 van Renselaer, S., *Industrial Fire Safety*, Jan./Feb. 1993, p. 32.
6 Alexander, J. M., *The Chemical Engineer*, No. 514, 27 Feb. 1992, p. 4.
7 Kletz, T. A., *Lessons from Disaster – How Organisations have No Memory and Accidents Recur*, Institution of Chemical Engineers, Rugby, UK, 1993.
8 Kletz, T. A., *Plant/Operations Progress*, Vol. 12, No. 3, July 1993, p. 47.

Chapter 7

Another tank explosion – The hazards of modification and ignorance

It happens, like as not,
There's an explosion and good-bye the pot!
These vessels are so violent when they split
Our very walls can scarce stand up to it. –

Geoffrey Chaucer (*c.* 1386),
The Canon Yeoman's Tale in *The Canterbury Tales*,
translated by Neville Coghill

One end of a horizontal cylindrical tank (volume 20 m³) was blown off killing two men who were working nearby (Figure 7.1). The tank was used to store a product of melting point 100°C which was handled as a liquid and kept hot by a steam coil. The gauge pressure of the steam was 100 psi (7 bar). The line into the tank was steam traced and insulated but any faults in the steam tracing or insulation resulted in the formation of plugs of solid; the line had had to be dismantled on several occasions to clear them. It was therefore emptied by blowing with compressed air after it had been used and it was again blown with compressed air, to prove that it was still clear, before it was used again.

The accident happened during this blowing operation. An operator connected an air hose to the transfer line at the production unit, about 200 m away, and then went to the tank to see if he could hear the air coming out of the vent – an open hole 3 inches diameter. The air was not getting through. He spoke to two maintenance workers who were working nearby on a job unconnected with the tank. He then went to report to his foreman that no air was coming out of the tank vent and that the transfer line was presumably choked, not for the first time! While he was talking to the foreman they heard a bang and going outside found that one end of the tank had blown off, killing the two maintenance workers.

It was later found that the vent was plugged with solidified product. The tank was designed to withstand a gauge pressure of 5 psi (0.3 bar) and would burst at a gauge pressure of about 20 psi (1.3 bar) so in retrospect it is not surprising that the plug of solid was stronger than the tank.

Figure 7.1 The end blown off the tank by gas pressure

In discussing this accident it will be convenient to deal with each aspect in turn – the control of plant modifications, the choking of the vent, knowledge of material strength and learning from the past – considering each at different depths.

7.1   The control of plant modifications

According to the design drawings a 6 inch diameter open hole was originally used as the tank vent. The plant had been in operation for about seven years and sometime during this period the 6 inch hole had been blanked off and an unused 3 inch opening on the tank used instead. (The tank was a second-hand one and was provided with more openings than were needed for its new duty.) No-one knew when the change had been made, who authorised it, or why. No records had been kept. It is possible that the 6 inch hole allowed dirt to enter the tank and contaminate the product. Experience elsewhere on the plant, which had many similar tanks and handled many similar products, showed that while a 3 inch hole could block with solid it was most unlikely that a 6 inch one would do so.

Blanking the 6 inch opening, which ended in a standard flange, and removing a blank from the 3 inch opening would have taken a fitter little more than an hour. The cost would have been negligible and no financial authorisation would have been required. The only paperwork needed would have been a permit-to-work which any of the shift foremen could have issued. Yet the modification led to the deaths of two men. The incident shows very clearly the results that can follow from unauthorised modifications and the need for a system to control them. References 1–4 and Chapters 1 and 8 describe other accidents that have resulted from modifications that had unforeseen side-effects and the first two references describe the essentials of a control system. This should consist of four parts:

(1) *Procedures* No modification should be made to plant or process unless it has been authorised in writing by a competent person, who should normally be professionally qualified. Before giving that authorisation the competent person should specify the change in detail and should try to identify all the consequences, discussing the modification with experts in mechanical and electrical engineering, control, chemical reaction hazards, materials, etc., as appropriate. The various experts should normally meet together. If not, the competent person should discuss the modification with them individually. Circulating the proposal is rarely satisfactory. When the modification is complete, and before it is commissioned, the competent person should inspect it to see that his intentions have been followed and that the modification 'looks right'. What does not look right is usually not right and should be checked thoroughly (Figure 1.6).

(2) *Aids* Substantial modifications should be subjected to a hazard and operability study (hazop)[5,6]. For minor modifications (for example, fitting an additional valve or sample point or making a small increase in the maximum permitted temperature or pressure) it is sufficient to complete a guide sheet or check list, designed to help people identify the consequences of the modification. One such guide sheet is shown in references 1, 2 and 5. The people completing the guide sheet should discuss it, not just pass it from one person to another.

(3) *Training* People will follow this procedure only if they are convinced that it is necessary. A training programme is needed. A discussion of accidents that have occurred as the result of ill-considered modifications is more effective than a lecture. Suitable slides and notes are available from the Institution of Chemical Engineers[7].

(4) *Monitoring* As with all safety procedures, regular auditing, to make sure that the system is being used, is necessary.

7.2   The vent was a relief valve

An open vent is in effect a relief valve. It is a very simple relief valve but it fulfils the function of a relief valve and it should be treated with the same respect, that is, it should be registered for regular inspection (to make sure that its size has not been changed and that it is not obstructed) and it should not be altered in any way without going through the same procedure as we go through before modifying a relief valve.

Unfortunately those concerned did not realise this. No process manager, foreman or operator would dream of asking a fitter to replace a relief valve by a smaller one unless calculation has shown that a smaller one is adequate and no fitter would carry out the replacement. Yet replacement of a 6-inch vent by a 3-inch one seems to have been done at the drop of a hat. Many companies still treat open vents with less respect than relief valves. Let us hope they learn from this incident and do not wait until a similar one occurs on their own plant.

As stated in Chapter 1, equipment whose size or existence has been taken into account in sizing a relief valve or vent, such as non-return (check) valves, control valve trims and restriction orifice plates, should also be registered for regular inspection and should not be modified or removed without checking that the size of the relief valve or vent will still be adequate.

As already stated the six inch hole was fitted with a standard flange, a survival from the vessel's previous duty. This made it too easy for the vent to be blanked. Vent openings should be designed so that they cannot easily be blanked.

7.3   The choking of the vent

Choking of the vent could have been prevented by retaining its original 6 inch size. It could also have been prevented by heating the vent (with steam or electric tracing) and/or by inspecting the vent to see if it was choked and rodding it clear if necessary. Some operators did this but the access was poor and others did not bother. After the accident it came to light that one of the shift foremen had complained to the manager several times about the difficult access and had asked for improvements to be made but nothing had been done.

Nothing had been done because the choking of the vent was looked upon, by all concerned, as an inconvenience rather than a hazard and improvements got the priority given to an inconvenience, not the priority given to a hazard. Nobody realised that the vessel was too weak to withstand the air pressure, if the vent was fully choked.

After the accident changes were made to the vent, and to the vents on 200 similar tanks. The vents were steam jacketed, fitted with flame traps (the reason for these is described later) and, in case the vents should choke, a steam-heated blow-off panel was provided (see Section 7.6). The access was also improved. Each vent was inspected every day to make sure that it was clear and that the steam heating was in operation. The frequency of inspection was much higher than for a normal relief valve, where once every two years is considered adequate, but was necessary as the vents can choke quickly if the steam supply fails. A special inspector had to be appointed as the volume of work involved was considered too much for the operators.

There may be many other operators, and managers, who do not know the strength of their equipment. All plant staff should know the pressures, temperatures, concentrations of corrosive materials, and so on that their equipment can safely withstand and the results that will follow if they go beyond them.

The operators found it hard to believe that a 'puff of air' had blown the end off the tank and explosion experts had to be brought in to convince them that a chemical explosion had not occurred. On many other occasions operators have been quite unaware of the power of compressed gases[8].

7.4   Mitigation of consequences

An ordinary atmospheric pressure storage tank would have been adequate for the storage of the product. A second-hand pressure vessel was used as it happened to be available. The designer probably thought that by using a stronger vessel than necessary he was providing increased safety. In fact he was not. If an atmospheric pressure tank had been used and its vent had choked, the roof would have blown off at a gauge pressure of about 24 inches of water (less than 1 psi or about 6 kPa), not 20 psi, and the energy released would have been correspondingly less, though still high enough to kill someone who was standing in the direction of failure. However, as atmospheric pressure storage tanks are vertical cylinders, normally with weak seam roofs which blow off if the tank is overpressured, it is unlikely that anyone will be in the way. In contrast, if the end comes off a horizontal cylinder, someone is more likely to be in the way.

Similarly, if vessels are exposed to fire, weaker vessels fail at lower pressures, with less disastrous consequences, than stronger vessels[9].

Designers should therefore:

- Remember than stronger does not always mean safer.
- Consider the way in which equipment will fail (see also Section 2.7, penultimate paragraph).

Alternatively, the tank could have been designed to withstand the full pressure of the compressed air, but this would be an expensive solution.

7.5   Could it have been a chemical explosion?

The product in the tank had a flash point of 120°C and was kept at about 150°C, so a flammable atmosphere was present in the tank. There was no evidence that a chemical explosion had occurred and the tank failed at exactly the time that it was calculated the pressure would have risen to the bursting pressure. Nevertheless it is not good practice to blow compressed air into a tank containing flammable vapour and after the incident nitrogen was used instead. As air might diffuse into the tank a flame arrester was fitted in the vent. Two hundred other tanks in the plant were treated similarly.

7.6   The flame trap assembly

Figures 7.2 and 7.3 illustrate the flame trap assembly that was fitted to 200 tanks, in place of the manhole cover, after the accident[10]. The device consists of a short length of steam-jacketed pipe into which a crimped ribbon flame arrester is fitted. A long handle on the flame arrester enables it to be removed, without the use of tools, and held up to the light for inspection. The inspector can also look through the vent hole into the tank to see that there is no obstruction. Adequate access must, of course, be provided. Should the flame arrester choke, or be unable to pass the required quantity of gas, a loose D-shaped lid lifts and relieves the pressure. A chain prevents the lid going into orbit. Steam tracing prevents the lid from being stuck in position by adhering solid.

The steam jacket should extend at least 3 inches (75 mm) into the tank to prevent a bridge of solid forming across the bottom. The flame arrester should be about 9 inches (225 mm) from the bottom of the vent and about 3–4 inches (75–100 mm) from the top.

Manufacturers of flame arresters can supply data on the gas flows through them at various pressure drops. For example, a 4.66 inch (118 mm) diameter flame arrester, 0.75 inch (20 mm) thick, will pass 700 m³/h of air at a pressure drop of 8 inches (200 mm) of water (2 kPa).

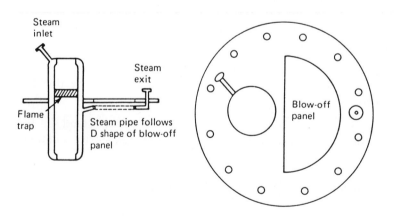

Figure 7.2 The flame trap assembly which was fitted in place of the man-hole cover. (a) Elevation section. (b) Plan from above.

Figure 7.3 The flame trap assembly, showing how the flame trap can be removed for inspection

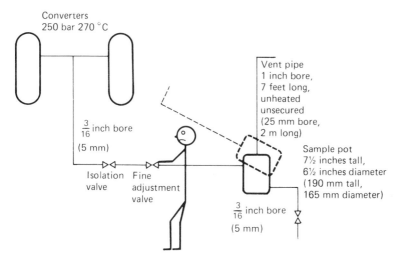

Figure 7.4 An attempt to clear a choke and obtain a sample from the line joining two reactors

7.7   Learning from the past

During the investigation someone recalled that a similar accident had occurred about 20 years before and after much searching the report was found. Another man had been killed by a choked vent about 100 m away although in a unit under the control of a different department.

A vegetable oil was being hydrogenated on a pilot plant and samples had to be taken from the line joining two reactors (Figure 7.4). The sample pot was filled by opening the isolation valve and cracking open the fine adjustment valve. Both valves were then closed and the sample drained into a sample can. As the oil and product had a melting point of 20°C, all lines were steam traced, except for the vent line.

Two weeks after start-up the operator was unable to get a sample. Suspecting a choke, his chargehand tried to clear it by opening the isolation and drain valves fully and then gradually opening the fine adjustment valve. The vent pipe moved suddenly and hit the chargehand on the head.

The sample pot and vent line had not been clamped and were free to rotate by partially unscrewing the flange on the inlet pipe. It had been intended to clamp them but this had been overlooked. It is believed that there was a choke somewhere in the system, probably in the vent pipe which was not heated, and that when it cleared the back pressure from the sudden rush of gas caused the vent pipe to move backwards.

The main recommendation of the report was that plants handling materials which are solid at atmospheric temperature should be adequately heated. This applied particularly to the vent pipe.

Many other examples could be quoted of serious accidents which have been forgotten, or not made known to other departments of the same organisation, and which have been repeated after 10 or more years[11]. Another was described in Chapter 4. (Less serious accidents are forgotten more quickly and repeated more often.) This will continue unless companies take action to keep memories alive and remind people of the lessons of past accidents. This is discussed further in Chapter 22 and in reference 11.

7.8   Coincidences

The tank which failed was used as the feed vessel for the packing building and was located on the roof of the building. It was not visited very often but unfortunately its end blew off at a time when two men were working nearby on other equipment. If a quantitative risk assessment had been carried out the chance of anyone being killed by the vessel bursting would have been estimated as one in hundred or less. Nevertheless two men were killed. This is the only accident in this book in which a true coincidence, as distinct from the phoney coincidences discussed in Section 2.7, played a part. On other occasions equally unlikely coincidences have

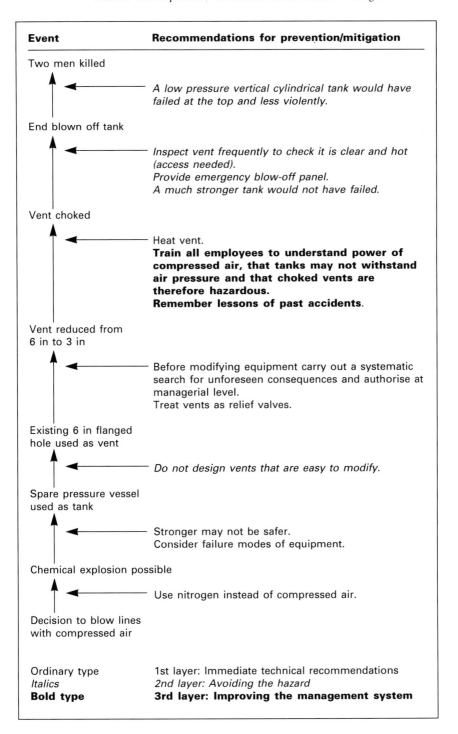

Event

Recommendations for prevention/mitigation

Two men killed

A low pressure vertical cylindrical tank would have failed at the top and less violently.

End blown off tank

Inspect vent frequently to check it is clear and hot (access needed).
Provide emergency blow-off panel.
A much stronger tank would not have failed.

Vent choked

Heat vent.
Train all employees to understand power of compressed air, that tanks may not withstand air pressure and that choked vents are therefore hazardous.
Remember lessons of past accidents.

Vent reduced from 6 in to 3 in

Before modifying equipment carry out a systematic search for unforeseen consequences and authorise at managerial level.
Treat vents as relief valves.

Existing 6 in flanged hole used as vent

Do not design vents that are easy to modify.

Spare pressure vessel used as tank

Stronger may not be safer.
Consider failure modes of equipment.

Chemical explosion possible

Use nitrogen instead of compressed air.

Decision to blow lines with compressed air

Ordinary type	1st layer: Immediate technical recommendations
Italics	*2nd layer: Avoiding the hazard*
Bold type	**3rd layer: Improving the management system**

Figure 7.5 Summary of Chapter 7 – Another tank explosion

prevented death or injury, for example, when someone has left the site of an explosion a few minutes before it occurred.

References

A brief account of this accident originally appeared in reference 10.

1 Kletz, T. A., *Chemical Engineering Progress*, Vol. 72, No. 11, Nov. 1976, p. 48.
2 Lees, F. P., *Loss Prevention in the Process Industries*, Vol. 2, Butterworths, 1980, Sections 21.6–21.8.
3 Kletz, T. A., *What Went Wrong? Case Histories of Process Plant Disasters*, 2nd edition, Gulf Publishing Co., Houston, Texas, 1988, Chapter 2.
4 Sanders, R. E., *Management of Change in Chemical Plants - Learning from Case Histories*, Butterworths-Heinemann, Oxford, 1993.
5 Kletz, T. A., *Hazop and Hazan - Identifying and Assessing Process Industry Hazards*, 3rd edition, Institution of Chemical Engineers, Rugby, UK, 1992.
6 Knowlton, R. E., *A Manual of Hazard and Operability Studies*, Chemetics International, Vancouver, Canada, 1992.
7 *Hazards of Plant Modifications – Hazard Workshop Module No 002*, Institution of Chemical Engineers, Rugby, UK, undated.
8 Kletz, T. A., *An Engineer's View of Human Error*, 2nd edition, Institution of Chemical Engineers, Rugby, UK, 1991, Section 2.2.
9 Kletz, T. A., *Improving Chemical Industry Practices – A New Look at Old Myths of the Chemical Industry*, Hemisphere, New York, 1990, Section 4.
10 Kletz, T. A., *Plant/Operations Progress*, Vol. 1, No. 4, Oct. 1982, p. 252.
11 Kletz, T. A., *Lessons from Disaster – How Organisations have No Memory and Accidents Recur*, Institution of Chemical Engineers, Rugby, UK, 1993.

Chapter 8

Flixborough

We often think, naively, that missing data are the primary impediments to intellectual progress – just find the right facts and all problems will disappear. But barriers are often deeper and more abstract in thought. We must have access to the right metaphor, not only to the requisite information.

S. J. Gould, *The Flamingo's Smile*

The explosion in the Nypro factory at Flixborough, England on 1 June 1974 was a milestone in the history of the chemical industry in the UK. The destruction of the plant in one almighty explosion, the death of 28 men on site and extensive damage and injuries, though no deaths, in the surrounding villages showed that the hazards of the chemical industry were greater than had been generally believed by the public at large. In response to public concern the government set up not only an enquiry into the immediate causes but also an 'Advisory Committee on Major Hazards' to consider the wider questions raised by the explosion. Their three reports[1] led to far-reaching changes in the procedures for the control of major industrial hazards. The long-term effects of the explosion thus extended far beyond the factory fence.

The plant on which the explosion occurred was part of a complex for the manufacture of nylon, jointly owned by Dutch State Mines (55%) and the UK National Coal Board (45%). It oxidised cyclohexane, a hydrocarbon similar to petrol in its physical properties, with air to a mixture of cyclohexanone and cyclohexanol, usually known as KA (ketone/alcohol) mixture. The reaction was slow and the conversion had to be kept low to avoid the production of unwanted by-products so the inventory in the plant was large, about 400 tonnes[2]. The reaction took place in the liquid phase in six reaction vessels, each holding about 20 tonnes. The liquid overflowed from one reactor to another while fresh air was added to each reactor (Figure 8.1). Unconverted raw material was recovered in a distillation section and recycled. Similar processes were operated, with variations in detail, in many plants throughout the world.

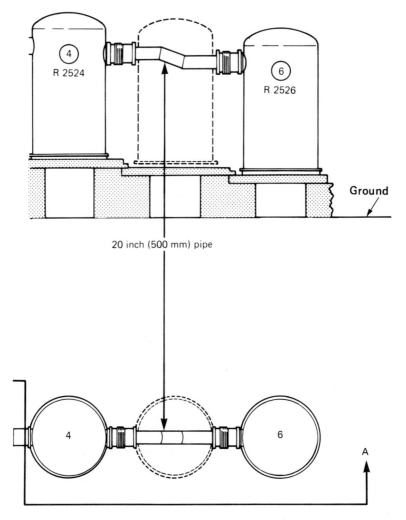

Figure 8.1 Arrangement of reactors and temporary pipe at Flixborough. (Reproduced by permission of the Controller of Her Majesty's Stationery Office)

One of the reactors developed a crack and was removed for repair. In order to maintain production a temporary pipe was installed in its place. Because the reactors were mounted on a sort of staircase, so that liquid would overflow from one to another, this temporary pipe was not straight but contained two bends (Figure 8.1). It was 20 inches diameter although the short pipes which normally joined the reactors together were 28 inches diameter. Calculation showed that 20 inches would be adequate for the flow rates required. Bellows, also 28 inches diameter, were installed between each reactor and these were left at each end of the temporary pipe.

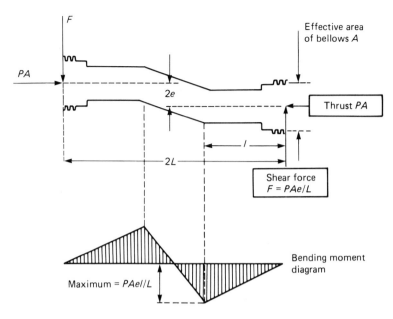

Figure 8.2 Sketch of pipe and bellows assembly showing shear forces on the bellows and bending moments in the pipe (due to internal pressure only). (Reproduced by permission of the Controller of Her Majesty's Stationery Office)

The men who were charged with the task of making and installing the temporary pipe were men of great practical experience and drive – they constructed and installed it and had the plant back on line in a few days. However, they were not professionally qualified and did not realise that the pipe and its supports should have been designed by an expert in piping design. Their only drawing was a full-size sketch in chalk on the workshop floor. The only support was a scaffolding structure on which the pipe rested.

The temporary pipe performed satisfactorily for two months until a slight rise in pressure occurred. The pressure was still well below the relief valve set point but nevertheless it caused the temporary pipe to twist (Figure 8.2). The bending moment was strong enough to tear the bellows and two 28 inch holes appeared in the plant. The cyclohexane was at a gauge pressure of about 10 bar (150 psi) and a temperature of about 150°C. It was thus under pressure above its normal boiling point (81°C) and a massive leak occurred as the pressure was reduced. It was estimated that about 30–50 tonnes escaped in the 50 seconds that elapsed before ignition occurred. The source of ignition was probably a furnace some distance away.

The resulting explosion, one of the worst vapour cloud explosions that has ever occurred, destroyed the oxidation unit and neighbouring units and caused extensive damage to the rest of the site. In particular the company office block, about 100 m away, was destroyed and had the explosion occurred during office hours, rather than at 5 pm on a Saturday, the death toll might have been 128 instead of twenty-eight. There is a detailed account of the explosion and of the events leading up to it in the official report[3] and in reference 4.

Instead of looking at the outer layers of the onion first and then the inner layers, it will be convenient to look at the various events leading up to the accident, summarised in Figure 8.4, in the order in which they occurred.

8.1   Reduction of inventory

One of the most important lessons to be learned from Flixborough, perhaps the most important, was ignored in the official report and in most of the other papers which have appeared on the subject: *If the inventory had been smaller the leak would have been less. What you don't have, can't leak*. In fact, it was Flixborough that made chemical engineers realise that it might be possible to reduce inventories. Before Flixborough most companies designed plants and accepted whatever inventory was called for by the design. Many companies still do so. However, an increasing number now realise that if we set out to reduce inventories it is often possible to do so and that the resulting plants are cheaper as well as safer. They are cheaper because less added-on protective equipment is needed, and has to be tested and maintained, and because a smaller inventory means smaller, and therefore cheaper, equipment. My book *Plant Design for Safety – A User-Friendly Approach*[5] gives many examples of what has been and might be done. Such plants are said to be inherently safer to distinguish them from conventional or extrinsically safe plants where the safety is obtained by adding on safety features.

Let us look at the scope for inventory reduction in the process operated at Flixborough. The output of the plant was about 50 000 tonnes/year KA. Assuming a linear velocity of 0.5 m/s, it could all have passed through a pipe 1.6 inches diameter. Any larger pipe is a measure of the inefficiency of the chemical engineering. Actual pipe sizes ranged up to 28 inches, so the cross-section of the pipe, and thus the flow rate, was $(28/1.6)^2 = 300$ times greater than the theoretical minimum. This is a bit unfair as the 28 inch pipes were made larger than necessary for the liquid flow so that the gas pressures in the reactors were the same. Photographs of the plant suggest that liquid lines were at least 12 inches diameter. The 'inefficiency' or 'hazard enhancement' index was thus $(12/1.6)^2 = 56$. Was it possible to reduce it?

The inefficiency was so large because reaction was slow and conversion low (about 6%). The reaction takes place in two stages, oxidation to a peroxide followed by its decomposition. The first stage is not really slow. Once the molecules come together they react quickly. It is the mixing that is slow, the chemical engineering rather than the chemistry, so if the mixing can be improved, reaction volume can be reduced. The second stage is slow and a long residence time is therefore required. If it could take place in a long thin tube instead of a vessel, leaks would be unlikely, would be restricted by the size of the tube and could be stopped by closing an emergency valve. Higher temperatures would increase the rate of decomposition.

Conversion was low because because if more air is used to increase conversion it is difficult to ensure that the air is intimately mixed with the hydrocarbon, and high local concentrations cause unwanted by-products to be formed. If a better method of mixing air and hydrocarbon can be found then conversion can be increased; one possible method would be to mix them under conditions in which reaction cannot occur and then change conditions so that reaction becomes possible.

Another problem that would have to be overcome in a new design would be cooling. The heat of reaction is considerable and is removed by evaporation and condensation of unconverted cyclohexane, and its return to the reactor. About 7 tonnes have to be evaporated for every tonne that reacts. A higher conversion process would require a large heat exchange area, a property of long thin tubes, unless some other method of energy removal, such as electricity generation in fuel cells, can be devised[6].

Of course, the company concerned, Nypro, cannot be blamed for adopting a high-inventory process. They were using much the same process as every other manufacturer of nylon and were following 'best current practice'. It simply had not occurred to the industry as a whole that inventory reduction was either possible or desirable. The chance of a major escape was considered so low that the large inventory was not looked upon as a hazard and everyone was so used to the process that its gross inefficiency was accepted without comment or complaint. Yet most of the material in the plant was getting a free ride; it was going round and round with only a small proportion being reacted and removed at each pass. What would we think of an airline which flew from London to New York and back with only 6% of the passengers alighting on each round trip, the rest staying on board to enjoy the movies?

Following Flixborough one company did devise a lower inventory process but abandoned the development programme when it became clear that no new plants would be needed in the foreseeable future. More recently, Union Carbide have described an improved pot reactor for gas/liquid reactions[7]. The gas is added to the vapour phase and is pulled down into the liquid by a down-pumping impeller which is used in place of a conventional stirrer. Higher gas solution rates are said to be obtainable without the risk of high local concentrations and the conversion is improved.

The Flixborough explosion thus demonstrates the desirability of reducing inventories. It also shows that the most important inventories to reduce are those of flashing flammable or toxic liquids, that is, liquids under pressure above their normal boiling points. Suppose a 2-inch hole appears in a line or vessel carrying a liquid such as petrol at a gauge pressure of 7 bar (100 psi) and at atmospheric temperature. Liquid will leak out at a rate of about 3 tonnes/minute but very little vapour will be formed (unless the liquid leaks at a high level and falls over the structure) and the chance of an explosion or large toxic cloud is very small. If the line or vessel contains a gas such as propane at the same pressure and temperature, the leak will be much smaller, about 1/4 tonne/minute, it may disperse by jet mixing and again the vapour cloud will not be large. However, if the line or vessel contains petrol at the same pressure but at a temperature of 100°C, above its normal boiling point, the liquid will again leak at a rate of about 3 tonnes/minute but will then turn to vapour and spray. The spray is just as explosive and just as toxic as the vapour and a large explosive or toxic cloud will be formed. Most unconfined vapour cloud explosions have resulted from leaks of flashing flammable liquids such as liquefied petroleum gas or hot hydrocarbon under pressure[8], and most toxic incidents from leaks of liquefied toxic gases such as chlorine or ammonia.

Since 1974 progress towards inherently safer plants, though significant, has been slower than was hoped. In part this has been due to the recession in the chemical industry – few new plants have been built – but there are also other reasons: our natural caution and conservatism when innovation is involved and the fact that inherently safer designs require more time during the early stages of design for alternatives to be evaluated and new ideas developed, time which is often not available if the plant is to be ready in time to meet a market opportunity. This applies to all innovation but particularly to innovation in the field of safety as in most companies safety studies do not take place and safety people do not get involved until late in design. If we are to develop inherently safer designs, we need to carry out studies similar to hazard and operability studies (hazops) much earlier in design, at the conceptual stage when we are deciding which process to use, and at the flowsheet stage, before detailed design starts. These studies should be additional to the usual hazop of the line diagram and will not replace it. In addition, Malpas[9] suggests that during design, one member of the design team should be responsible for thinking about the plant after next, as during design we are conscious of all the changes we would like to make but cannot make as there is insufficient time. These constraints on the development of inherently safer designs are discussed in more detail in reference 5.

8.2   The control of process modifications

Why did a crack appear in one of the reactors? There was a leak of cyclohexane from the stirrer gland and, to condense the leaking vapour, water was poured over the top of the reactor. Plant cooling water was used as it was conveniently available.

Unfortunately the water contained nitrates and they caused stress corrosion cracking of the mild steel pressure vessel (though fortunately not of the stainless steel liner). Nitrate-induced cracking was well known to metallurgists but was not well known, in fact, hardly known at all, to other engineers[10].

The next section of this chapter and the explosion described in Section 7.1 show the importance of controlling plant modifications. Process modifications, as this crack shows, can also produce unforeseen results. No change in operating conditions, outside the approved range, should be made without going through the procedure described in Section 7.1, that is, they should not be made until they have been authorised by a professionally qualified manager who should first use a systematic technique to help him identify the consequences. Pouring water over equipment was at one time a common way of providing extra cooling but it is nevertheless a process change that should go through the modification procedure. *The more innocuous a modification appears to be, the further its influence seems to extend.*

However, no great blame can be attached to the Nypro staff. They did what many people would have done at the time. The purpose of investigations, and of this book, is not to find culprits but to prevent future accidents.

8.3   The control of plant modifications

As already described, the temporary pipe was installed very quickly, to restore production, and there was no systematic consideration of the hazards or consequences. One of the main lessons to be learned is that stated in Section 7.1, that no modification should be made until it has been authorised in writing by a competent person who should normally be professionally qualified, and that before giving that authorisation the competent person should try to identify all the consequences and should specify the change in detail. When the modification is complete and before it is commissioned he should inspect it to see that his intentions have been followed and that the modification 'looks right'. In particular, the modification should be made to the same standard as the original design and this standard should be specified in detail. In addition there should be a guide sheet or check list to help the competent person identify the consequences and a training programme is needed to convince people that the modification control procedure is necessary. As with all procedures, regular monitoring is needed to make sure that the procedure is being followed.

Note that modifications to supports can be as important as modifications to equipment and that if pipes are being modified the supports need as much consideration as the pipe itself.

8.4   The need for suitably qualified staff

The reference to professional qualifications is important. The men who constructed the temporary pipe did not know how to design large pipes capable of withstanding high temperatures and pressures. Few engineers do. It is a specialised branch of mechanical engineering. However, a professional engineer would have recognised the need to call in an expert in piping design. The men who constructed the pipe did not even know that expert knowledge was needed: *They did not know what they did not know.* As a result they produced a pipe that was quite incapable of withstanding the operating conditions. In particular, to install an unrestrained pipe between two bellows was a major error, specifically forbidden in the bellows manufacturer's literature.

At the time the pipe was constructed and installed there was no professionally qualified mechanical engineer on site, though there were many chemical engineers. The establishment called for one professional mechanical engineer, the works engineer, but he had left and his successor, though appointed, had not yet arrived. Arrangements had been made for a senior engineer of the National Coal Board, who owned 45% of the plant, to be available for consultation, but the men who built the pipe did not see the need to consult him. Another lesson of Flixborough is, therefore, the need to see that the plant staff are a balanced team, containing people of the necessary professional experience and expertise. On a chemical plant mechanical engineering is as important as chemistry or chemical engineering. To quote a former chief engineer of ICI's Billingham Division, *A place like Billingham is really engineering; chemistry is only what goes through the pipes.*[11]

Since Flixborough most chemical plants have reduced staff but the need for professional expertise remains as great as ever. For example, a plant which at one time employed an electrical engineer may no longer do so, the control engineer having to act as electrical engineer as well. There is an electrical engineer available for consultation somewhere in the company, but will the control engineer know when to consult him? Will he know what he does not know?

Should the chemical engineers have noticed that the temporary pipe did not look right? At the time some of my colleagues said that if they had been on the plant they would have questioned its design. Others thought that it was unreasonable to expect a chemical engineer to do so and certainly those who were on the plant cannot be blamed for not noticing mechanical design errors. The official report[3] suggested that the training of engineers should be more broadly based. 'Although it may well be that the occasion to use such knowledge will not arise in acute form until an engineer has to take executive responsibility it is impossible at the training stage to know who will achieve such a position' (paragraph 210).

The explosion does also show the importance of plant staff, at all levels from the most senior downwards, spending part of each day out on the plant looking round. A plant cannot be managed from an office. Unfortunately, too many managers find that the volume of paperwork provides a good excuse for staying in the warmth and comfort of their offices.

If a chemical engineer had felt that the appearance of the temporary pipe was not right he might have felt that it was none of his business. Perhaps the mechanical engineers would resent his interference. Presumably they know their job. Better say nothing. Flixborough shows us that we should never hesitate to speak out if we suspect that something may be wrong.

In the light of Flixborough, companies should ask themselves not only if they have enough mechanical engineers, electrical engineers and so on, but also if their safety advisers have the right technical knowledge and experience and sufficient status for their opinions to carry weight. In high technology industries the safety adviser needs to be qualified and experienced in the technology of the industry, with previous line management experience, or the managers will not listen to him. He needs to be a numerate man, able to assess risks systematically and numerically, and a good communicator[12].

8.5   Robust equipment preferred

No criticism can be made of the manufacturers of the bellows. They were misused in a way specifically forbidden in their literature. Nevertheless bellows are inevitably a weak link, they will not withstand poor installation to the same extent as fixed pipework, and they should not be used when hazardous materials are handled. Instead expansion loops should be built into the pipework.

Generalising, we should, whenever possible, avoid equipment which will not withstand mistreatment. Flexible hoses are another example. To use a computer phrase, whenever possible we should use equipment which is 'user-friendly'[5].

8.6   Plant layout and location

It is almost impossible to prevent ignition of a leak the size of that which occurred at Flixborough. (Not all such leaks ignite, but it is often a matter of chance which ones do or do not.) The leak may spread until it reaches a source of ignition. However, it is possible to locate and lay out a plant so that injuries and damage are minimised if an explosion occurs.

Most of the men who were killed were in the control room and were killed by the collapse of the building on them. Since Flixborough many companies have built blast-resistant control rooms, able to withstand a peak reflected pressure of 0.7 bar (10 psi). They are not designed to withstand repeated explosions but to withstand one explosion. They may be damaged but should not collapse, thus protecting the men inside and also preserving the plant records which may help the subsequent investigation. Reference 13 gives a design procedure.

Is it right to protect the men in the control room and ignore those who are outside? At Flixborough most of the those outside survived. Men in an ordinary unreinforced building are at greater risk than those outside.

The control room has to be near the plant but other occupied buildings can usually be placed further away so that they need not be strengthened. Office blocks and other buildings containing large numbers of people should certainly never be located close to the plant, as at Flixborough. If, for any reason, an occupied building has to be near the plant, then it should be strengthened. How near is near? To answer this question, I suggest the following procedure[14]:

(1) Estimate the largest leak that should be considered, ignoring vessel failures, taking pipe failures into account but also taking credit for remotely operated emergency isolation valves.
(2) Estimate the size of the vapour cloud, by doubling the amount that, theoretically, will flash (unless the theoretical flash is over 50%) to allow for spray.
(3) Estimate the pressure developed at various distances if this cloud ignites. It is necessary to allow for the efficiency of the explosion and I suggest a figure of 2%. However, some writers have suggested higher figures.
(4) Design the building to withstand the pressure developed at the point at which it is placed.

The same procedure can be used to find a satisfactory layout and location for the plant. Thus other plants containing hazardous materials should not be placed at points at which the peak incident pressure will be 0.35 bar (5 psi) or more or they may be so seriously damaged that a further leak and a further explosion occur (the so-called domino effect). For low pressure storage tanks containing hazardous materials the maximum acceptable pressure is 0.2 bar (3 psi). Public roads should not be located where the pressure exceeds 0.07 bar (1 psi) and houses where it exceeds 0.05 bar (0.7 psi), preferably 0.03 bar (0.4 psi). The conclusions are summarised in Figure 8.3 and reference 14 gives more detail.

Figure 8.3 shows that there should be no buildings at all within 20 m of plant containing materials which might explode. This is to provide a fire barrier. It is good practice to divide large plants into blocks, separated by gaps about 15–20 m wide.

Some writers have suggested lower pressures, some higher. However, they have also suggested different methods for the calculation of the pressure, in particular different explosion efficiencies, and their final conclusions are not much different from those shown in Figure 8.3. Each method of estimation should be considered as a whole and we should not pick and choose figures from different methods.

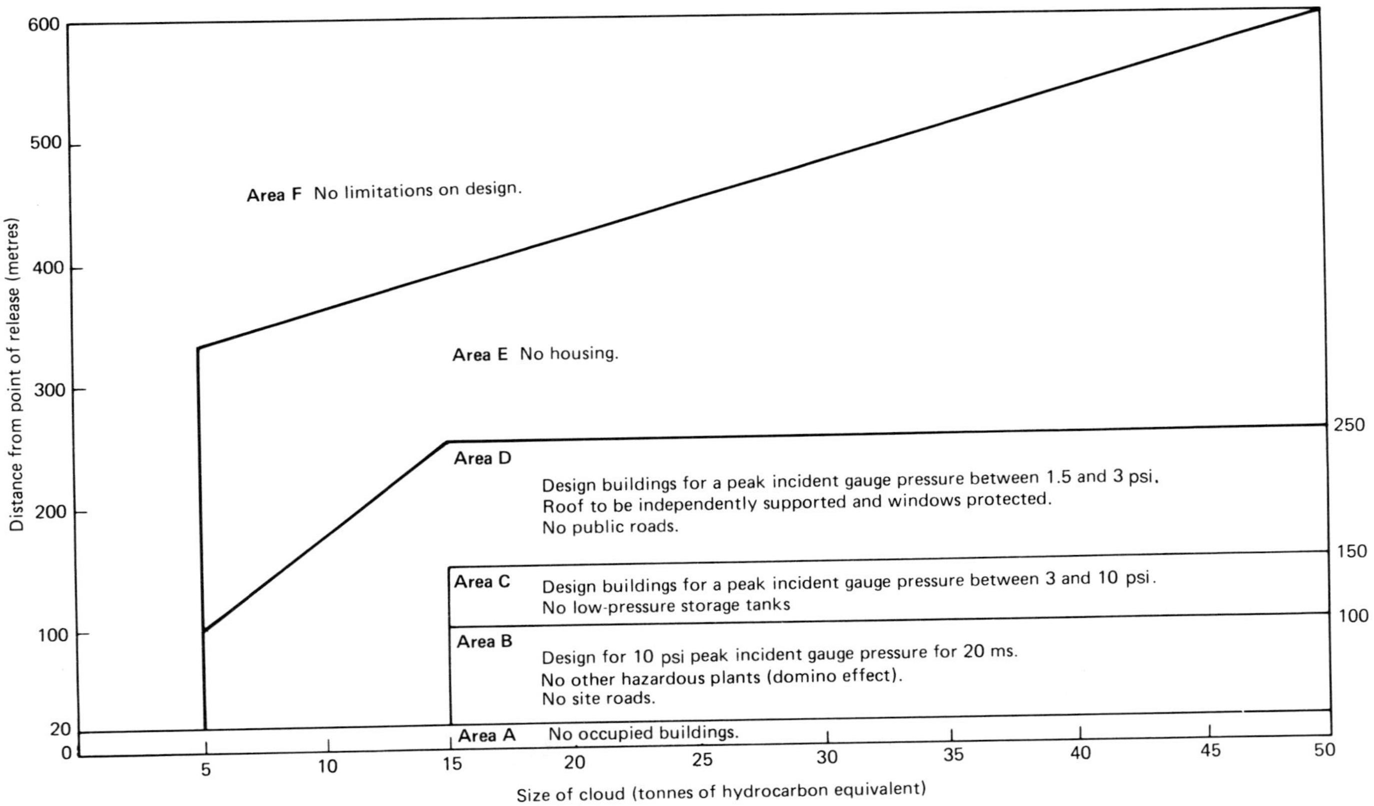

Area F No limitations on design.
Area E No housing.
Area D
Design buildings for a peak incident gauge pressure between 1.5 and 3 psi.
Roof to be independently supported and windows protected.
No public roads.
Area C
Design buildings for a peak incident gauge pressure between 3 and 10 psi.
No low-pressure storage tanks
Area B
Design for 10 psi peak incident gauge pressure for 20 ms.
No other hazardous plants (domino effect).
No site roads.
Area A
No occupied buildings.
Distance from point of release (metres)
600
500
400
300
200
100
20
0
250
150
100
5
10
15
20
25
30
35
40
45
50
Size of cloud (tonnes of hydrocarbon equivalent)

Figure 8.3 Summary of the main restrictions on design imposed by unconfined vapour cloud explosions. (Note: Area A limitations apply in areas E–A and so on)

A few companies have questioned the need for the precautions described above. It would be better, they say, to prevent explosions than guard against the consequences. This is true and we do try to prevent explosions, but at the present time experience shows that the probability of an explosion is not so low that it can be ignored.

A final point to be considered under layout is the location of the assembly points at which people congregate when a warning sounds and a plant is evacuated. They should be some distance away, at least 150 m, from equipment containing large inventories, in Area D of Figure 8.3. Control rooms should not be used as assembly points.

Layout and location are considered further in Section 10.2 and in reference 15.

8.7   Prevention of leaks

The Flixborough explosion was followed by an explosion of papers on the probability of large leaks and on their behaviour: How far will they spread various wind and weather conditions? What pressures will be developed if they ignite? And so on. In contrast, there has been very little written on the reasons for other large leaks and the actions needed to prevent them. It is as if the occasional large leak is considered inevitable.

Chapter 16 will show that about half the large leaks that have occurred have been the result of pipe failures and that many pipe failures have occurred because construction teams did not follow the design in detail or did not do well what was left to their discretion. The most effective single action we can take to prevent major leaks is therefore to specify piping designs in detail and then to check during and after construction that the design has been followed and that items not specified in the design have been constructed in accordance with good engineering practice. Much more thorough checking is needed than has been customary in the past. The construction inspector should look out for errors that no one would dream of specifically forbidding.

8.8   Public response

The material damage at Flixborough cost the insurance companies about £50million (equivalent to well over £300million [$450million] at 1994 prices). It cost the industry many times more in the extra equipment they were required to install and the extra procedures they were required to carry out. Much of this expenditure should have been undertaken in any case. Some of it went further than was really necessary and was the result of the understandable public reaction to the explosion. It demonstrated that in the eyes of the public the industry is one, and that the industry as a whole is held responsible for, and suffers for, the shortcomings of individual companies. It therefore shows that companies should freely exchange information on accidents that have happened and the action needed to prevent them happening again. Unfortunately companies are now less willing to exchange information than they were in the 1970s. Reference 16 discusses the reasons in detail. The most important reasons are time – with reductions in numbers managers have less time to do the things they know they ought to do but do not have to do by a specific date – and, particularly in the USA, the influence of company lawyers who fear that disclosure may influence claims for damages or prosecution.

In the UK the government, as already stated, set up an Advisory Committee on Major Hazards to consider the wider implications of the Flixborough explosion. It took about ten years for their recommendations to be made and to come into force and they resulted in the CIMAH (Control of Industrial Major Accident Hazards) Regulations[17]. These require companies which have more than defined quantities of hazardous materials in storage or process to prepare a 'safety case' describing the action they have taken to identify and control the hazards. Two years after Flixborough the Seveso accident (Chapter 9) caused a similar public reaction across Europe and resulted in the so-called Seveso Directive of the European Community[18]. The CIMAH Regulations, as well as being the UK's response to Flixborough, also brought the Seveso Directive into force in the UK. In the US there was no similar response until after Bhopal (Chapter 10).

The CIMAH Regulations were made under the Health and Safety at Work Act and like the Act are 'inductive', that is, they state an objective that has to be achieved but do not say how it must be achieved, though guidance is available. In some countries, including the US, it seems to be believed that industrial accidents can be prevented by the government writing a book of regulations that looks like a telephone directory, though it is rather less interesting to read. In other countries, including the UK, it has been realised that it is impracticable to control complex, developing, high-technology industries by detailed regulations. They soon get out-of-date or circumstances arise which were not thought of by those who wrote them and they encourage the belief that all you need to do to get a safe plant is to follow the rules. Under the UK Health and Safety at Work Act, passed by Parliament at the time of Flixborough but not then fully in force, there is a general obligation to provide a safe plant and system of work and adequate instruction, training and supervision, so far as is reasonably practicable. In the first place it is up to the employer to decide what is a safe plant, etc., but if the Health and Safety Executive (HSE) – the body responsible for enforcing the Act – do not agree they will say so and if necessary issue an improvement or prohibition notice.

The UK approach is summed up in the following extract from an article entitled *Can HSE Prevent another Flixborough?* written by a member of the HSE[19]: 'On his own, the inspector is unlikely ever to be able to provide adequate safeguards or deterrents. He must and should rely heavily on industry where the expertise in a particular field is surely to be found. He must however learn the know-how necessary to identify and follow up weaknesses in both management and systems.'

Nevertheless, many people in the chemical industry feel that the public and governmental reaction to Flixborough has been excessive; they contrast it with the relative indifference to some other industrial hazards such as those of the construction industry, and non-industrial hazards such as the roads. Perhaps the industry needs to do more to explain the true size of the hazards and, equally important, the benefit the public get back in return. The public accept a terrible death toll on the roads because the advantages of the motor car are clear and obvious to all. In contrast the chemical industry is seen as making unpleasant chemicals with unpronounceable names in order to increase its sordid profits. At the best it provides exports and employment. The public does not realise that it provides the necessities for our modern way of life.

The Flixborough explosion affected the insurance companies as well as the public. It made many of them realise that individual claims might be larger than they had previously foreseen and many of them increased their premiums. It also made them ask themselves if they were doing enough to distinguish between good and bad risks and many of them increased their investment in systematic methods of risk assessment, thus accelerating a trend that was already under way. It is, of course, relatively easy for them to recognise those plants which are better designed. It is much more difficult to distinguish between good and bad management but there is growing interest in methods of making a quantitative assessment of management quality[20-22].

8.9   The responsibilities of partnerships

The Flixborough plant was jointly owned by Dutch State Mines, who supplied the know-how, and the UK National Coal Board. In such joint ventures it is important to be clear who is responsible for safety in both design and operation. Shortcomings in this respect do not seem to have been responsible in any way for the Flixborough explosion but they were relevant at Bhopal (Chapter 10).

8.10   The rebuilt plant

When the plant was rebuilt the product, cyclohexanol, was manufactured by a safer route, much to the relief of the local population. Instead of manufacturing it by the oxidation of cyclohexane it was made by the hydrogenation of phenol. It is doubtful if the company would have been allowed to rebuild the plant if they had wished to use the original process.

However, phenol itself has to be manufactured, usually by the oxidation of cumene to cumene hydroperoxide and its 'cleavage' to phenol and acetone. This process, judged by its record[23], is at least as hazardous, perhaps more hazardous, than the oxidation of cyclohexane. It was not carried out at Flixborough but elsewhere. The hazards were not really reduced, only exported.

The rebuilt plant had a short life. It was closed down, after a few years, for commercial reasons.

8.11   Progress in implementing recommendations

This has been a long chapter so in conclusion it may be useful to summarise the progress made in carrying out the various recommendations made after Flixborough.

(1) On some issues progress has been good. Much more attention is now paid to plant layout and location and the strengthening of control buildings, and guidance is available for the designer. Many companies have established procedures for the control of modifications. Insurance companies play a more active role in encouraging good design. The competence of safety advisers has improved.

(2) In contrast, progress in the development of inherently safer designs for cyclohexane oxidation has been slower than expected, partly, perhaps, because all the new KA plants built since Flixborough have been in Eastern Europe or in newly developed countries. Little attention has been paid to the reasons for leaks and ways of preventing them.

Progress in these areas is important because while we can control small leaks by the methods discussed in Chapter 4 – emergency isolation valves, steam and water curtains, open construction and so on – we cannot control leaks of Flixborough size. Once a leak of this size occurs, all we can do is hope that it will not ignite (and, if time permits, evacuate the plant).

(3) In other areas progress varies greatly from company to company or even between different plants belonging to the same company. Many managers spend much of their time out on the plant while others do not.

(4) In some countries governments are seen as playing a helpful role, at least by the more safety conscious sections of industry. In others it is questioned whether their activities result in a net increase in safety. The public generally have a poor understanding of the chemical industry's contribution to our way of life and of its safety record.

Appendix: Flixborough and the Tay Bridge

In 1879 the Tay Bridge in Scotland collapsed six months after it was opened and a train fell into the river; all the passengers and crew – 75

Events	Recommendations for prevention/mitigation

New regulations and public
concern compelled other
companies to improve
standards and demonstrate
that they are safe

Provide information which will help politicians and public keep risks in perspective.

Destruction of plant
and death of 28 men.
Damage and injuries
outside factory.

Strengthen control buildings. Lay out and locate plants to minimise effects of explosions. Locate assembly points outside plant area.

Explosion

[Impossible to prevent ignition of a leak of this size.]

Leak of 30–50 tonnes
hydrocarbon

[Impossible to isolate or control a leak of this size.].

Pipe collapses

Provide adequate supports. Avoid bellows.

Slight rise in pressure

Temporary pipe
in use for 3 months

Look at plants with critical eyes on routine tours.

Reactor replaced by
temporary pipe

Construct modifications to same standard as original design. Employ professional mechanical engineers.

Crack in reactor

Cooling water poured on
reactor to control leak

Before modifying process carry out a systematic search for unforeseen consequences and authorise at managerial level.

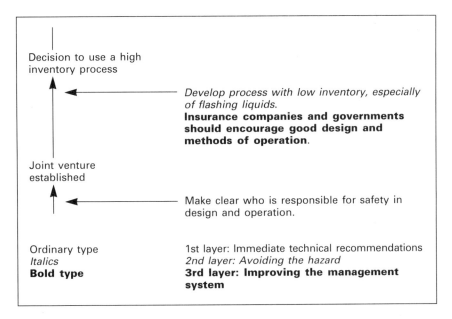

Figure 8.4 Summary of Chapter 8 – Flixborough

people – were drowned. It was not the worst British railway accident but it is the best remembered. The failure of a major engineering work so soon after completion caused considerable disquiet among both the public and professional engineers and, as at Flixborough, there was a public inquiry. The report shows a number of similarities to the Flixborough explosion.

(1) The failure of the Tay Bridge was due to a change in the original design. It was intended to construct the piers entirely of brick and concrete and fourteen out of fifty were made this way. However, during construction the river bed was found to be less strong than expected and so, to reduce the pressure on the river bed (and also to save cost), the designer, Sir Thomas Bouch, decided to build the piers of brick and concrete up to the high water mark and then construct the upper portions from cast iron pipes with cross-bracing. The brick and concrete piers still stand. Many of the original girders which went across the tops of the piers were re-used and are still in use today, but the cast iron pipes collapsed after six months (and did not turn out to be cheaper).

(2) The casting and erection of the iron pipes were not adequately super-vised. As a result the quality of the iron was poor, its thickness was uneven and many of the lugs to which the cross-bracing was attached were not secure. The subsequent inspection of the bridge was entrusted to a man who was very competent in his own field,

bricklaying, but lacked *relevant* experience. To quote from *The Tay Bridge Disaster* by John Thomas[24]:

The most embarrassing of all the NB (North British Railway) witnesses to the company was Henry Noble, the most honest and competent of men in his own limited sphere, but a bricklayer and not a man many railway companies would have chosen to take charge of the bridge.'

One of the lawyers wrote:

Mr Noble, as you know, is not a man of skill as regards ironwork. He is a good bricklayer. That is all. Yet he, and men much more ignorant than he, were apparently left to look after the ironwork of the bridge. No man of skill apparently went over it from week to week, or month to month. This point I think might be pressed home against the company very much'.

(3) The publication of the official report was followed by a debate on the extent to which reliance should be placed on official approval and inspection. Many people thought that government inspectors should have supervised the project so closely that they could guarantee its safety. On the other hand the President of the Board of Trade, Joseph Chamberlain, wrote (in a minute presented to Parliament):

If any public department were entrusted with the power and duty of correcting and guaranteeing the designs of the engineers who are responsible for railway structures, the result would be to check and control the enterprise which has done so much for the Country, and to substitute for the real responsibility which rests upon the railway engineer the unreal and delusive responsibility of a public office.

While the Flixborough Inquiry concluded that 'The blame for the defects in the design, support and testing of the by-pass must be shared between the many individuals concerned, at and below Board level' (paragraph 225), the Chairman of the Tay Bridge Inquiry blamed Sir Thomas Bouch entirely. (The other two members of the Inquiry did not believe they had to blame anyone though they agreed that the bridge was badly designed, constructed and maintained.) Many people felt that the Chairman was unfair and that many other people shared the responsibility for the accident, including the Board of Trade inspector who had inspected the bridge before giving permission for it to be used.

Sir Thomas Bouch's reputation was finished; his health deteriorated and he died soon afterwards. He had made his reputation by building cheaper bridges than anyone had built before, but in the Tay Bridge he went too far, in economising in the inspection of the ironwork during both construction and operation.

The wind, on the night of the collapse, was strong but no stronger than assumed during design. The high wind, like the slight rise in pressure in the temporary pipe at Flixborough, was a triggering event, not a cause.

References

1 Advisory Committee on Major Hazards, *First, Second and Third Reports*, Her Majesty's Stationery Office, London, 1976, 1979 and 1984.

2 Lewis, D. J., *The Chemical Engineer*, No. 462, July 1989, p. 4.

3 Parker, R. J., (Chairman), *The Flixborough Disaster: Report of the Court of Inquiry*, Her Majesty's Stationery Office, London, 1975.

4 Lees, F. P., *Loss Prevention in the Process Industries*, Vol. 2, Butterworths, 1980, Appendix 1.

5 Kletz, T. A., *Plant Design for Safety – A User-Friendly Approach*, Hemisphere, New York, 1991.

6 Ramshaw, C., Private communication.

7 Litz, L. M., *Chemical Engineering Progress*, Vol. 81, No. 11, Nov. 1985, p. 30.

8 Kletz, T. A., *The Chemical Engineer* No. 426, June 1986, p. 62.

9 Malpas, R., in *Research and Innovation for the 1990's*, edited by B. Atkinson, Institution of Chemical Engineers, Rugby, UK, 1986, p. 26.

10 *Guide Notes on the Safe Use of Stainless Steel in Chemical Process Plant*, Institution of Chemical Engineers, Rugby, UK, 1978.

11 Mayne, P., about 1950, quoted in *Chemical Age*, 20 Sept. 1974, p. 21.

12 Kletz, T. A., 'The safety adviser in the high technology industries – Training and role', *Safety and Loss Prevention*, Symposium Series No 73, Institution of Chemical Engineers, Rugby, UK, 1982, p. B1.

13 *An Approach to the Categorization of Process Plant Hazards and Control Building Design*, Chemical Industries Association, London, 1979.

14 Kletz, T. A., *Loss Prevention*, Vol. 13, 1980, p. 147.

15 Mecklenburgh, J. C. (ed.), *Process Plant Layout*, Godwin, London, 1985.

16 Kletz, T. A., *Lessons from Disaster – How Organisations have No Memory and Accidents Recur*, Institution of Chemical Engineers, Rugby, UK, 1993, Chapter 5.

17 *Control of Industrial Major Accident Hazard Regulations*, Statutory Instrument No 1902, Her Majesty's Stationery Office, London, 1984. See also *A Guide to the Control of Industrial Major Accident Hazard Regulations 1984*, Her Majesty's Stationery Office, London, 1990.

18 European Community: Council Directive of 24 June 1982 on the Major Accident Hazards of Certain Industrial Activities, *Official Journal of the European Communities*, 1989, No. L230, 5 August 1982, pp. 1–18.

19 Offord, D. V., in *Her Majesty's Inspectors of Factories: Essays to Commemorate 150 Years of Health and Safety Inspection*, Her Majesty's Stationery Office, London, 1983, p. 57. This book gives a good account of the way in which the UK Factory Inspectorate, part of the Health and Safety Executive, goes about its work.

20 Hurst, N. W. *et al.*, *Journal of Hazardous Materials*, Vol. 26, 1991, p. 159.

21 Williams J. C. and Hurst N. W., 'A comparative study of the management effectiveness of two technically similar major hazard sites', in *Major Hazards Onshore and Offshore*, Institution of Chemical Engineers Symposium Series No. 130, 1992, p. 73.

22 Hurst, N. W., Bellamy, L.J. and Wright, M.S. 'Research models of safety management of onshore major hazards', in *Major Hazards Onshore and Offshore*, Institution of Chemical Engineers Symposium Series No. 130, 1992, p. 129.

23 Kletz, T. A., *Plant/Operations Progress*, Vol. 7, No. 4, Oct 1988, p. 226.

24 Thomas, J., *The Tay Bridge Disaster*, David and Charles, Newton Abbot, UK, 1972.

Chapter 9

Seveso

I ought to have known ... the possibility of Singapore having no landward defences, no more entered my mind than that of a battleship being launched without a bottom.

Winston Churchill[1].

At Seveso near Milan in Italy, on 10 July 1976, a discharge containing highly toxic dioxin contaminated a neighbouring village over a period of about 20 minutes. About 250 people developed the skin disease chloracne and about 450 were burned by caustic soda. A large area of land, about 17 km^2, was contaminated and about 4 km^2 was made uninhabitable. Although no one was killed it has become one of the best known of all chemical plant accidents. It led to the enactment by the European Community of the Seveso Directive[2], which requires all companies which handle more than defined quantities of hazardous materials to demonstrate that they are capable of handling them safely. In the UK the Seveso Directive was implemented in the CIMAH (Control of Industrial Major Accident Hazard) Regulations[3]. No accident has resulted in greater 'legislative fall-out' than Seveso, although in the UK the CIMAH Regulations owe their origin to Flixborough and would have been enacted, in a slightly different form, if Seveso had never occurred.

The discharge came from a rupture disc on a batch plant for the manufacture of 2,4,5–trichlorophenol (TCP) from 1,2,4,5–tetrachlorobenzene and caustic soda, in the presence of ethylene glycol. TCP is used to make the herbicide 245T (2,4,5–trichlorophenoxyacetic acid). Dioxin (actually 2,3,7,8–tetrachlorodibenzodioxin or TCDD) is not normally formed, except in minute amounts, but the reactor got too hot, a runaway reaction occurred, dioxin was formed and a rise in pressure caused the rupture disc to burst. The propelling gas was probably hydrogen formed in the runaway reaction. There was no catchpot to collect the discharge and the contents of the reactor, about 6 tonnes, including about 1 kg of dioxin, were distributed over the surrounding area.

A foreman heard the noise produced by the discharge and opened the cooling water supply to the reactor coils. If he had not done so the discharge would have been greater.

9.1   How did the reactor get too hot?

Italian Law required the plant to shut down for the weekend, even though it was in the middle of a batch. On the weekend of the accident the operator had to shut it down at a stage in the batch, the end of the reaction but before all the ethylene glycol has been removed by vacuum distillation, at which it had not been shut down before. Perhaps he had got behindhand with his duties. However, he had no reason to believe that it would be hazardous to shut it down at this stage. The reaction mixture was at 158°C, well below the temperature at which an exothermic reaction can start, believed at the time to be 230°C but possibly as low as 180°C. It was at first difficult to see how the reactor got hot enough for a runaway to start.

The reactor was heated by an external steam coil (Figure 9.1) which used exhaust steam, at a pressure of 12 bar and a temperature of about 190°C, from a turbine on another unit. The turbine was on reduced load, as units were shutting down for the weekend, and the temperature of the exhaust steam rose to about 300°C[4].

The temperature of the liquid could not get above its boiling point at the operating pressure, about 160°C, and so below the liquid level there was a temperature gradient through the reactor wall, the outside being at 300°C and the inside at 160°C. Above the liquid level the wall was at 300°C right through. When the steam was isolated the wall below the liquid level soon fell to the temperature of the liquid, but above the liquid level it remained hotter. Heat passed from the upper wall to the surface of the liquid. When, 15 minutes later, the stirrer was switched off, the

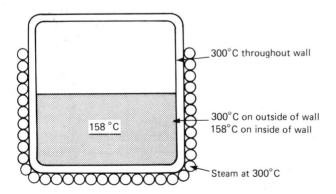

300°C throughout wall

300°C on outside of wall
158°C on inside of wall

158 °C

Steam at 300°C

Figure 9.1 The Seveso reactor. The hot upper portion of the reactor wall heated the surface of the liquid

upper few centimetres of the liquid rose in temperature from 158°C to 180–190°C and a slow exothermic reaction started; after about seven hours a runaway occurred[5-7]. It is possible that the rise in temperature was catalysed by material which had been stuck to the upper wall of the reactor, was degraded by the heat and fell into the liquid.

The runaway would not have occurred if:

- well-meaning legislators had not passed laws which left the management of the plant with no freedom to complete a batch before the weekend;
- the batch was not stopped at an unusual stage; or
- a hazard and operability study (hazop) had been carried out on the design. Provided the service lines were studied as well as the process lines, under the heading, 'more of temperature' this would probably have disclosed that when the turbine was on low load the temperature of the steam supplied to the reactor would rise. It is a well-established principle that, whenever possible, a heating medium should not be so hot that it can dangerously overheat the material being heated, and it was said to be a safety feature of the process used that there was a safety margin of 40°C between the temperature of the steam (192°C) and the runaway temperature (believed to be 230°C).

It is, of course, necessary when carrying out a hazop on a batch plant to consider each deviation, such as more of temperature, for each stage of the batch. All hazops should include service lines as well as process lines.

During a hazop we ask if the operator will know if a deviation occurs. (This is important for operability as well as safety.) Asking the question would have drawn attention to the fact that the temperature of the steam was not measured and a temperature measurement could have been added. In general, for every variable which can effect safety or operability significantly we need a measurement. On many plants the temperature of the steam cannot vary significantly or it does not matter if it does. If it can vary significantly, and it does matter, it should be measured.

9.2   Why was there no relief system catchpot?

When hazardous materials may be discharged from relief devices or vents it is normal practice to install a collection system: a flare system for flammable gases and vapours, a scrubbing system for toxic gases and vapours or a catchpot for solids and high-boiling liquids. Often more than one of these devices has to be provided. If a catchpot had been installed at Seveso the name would still be unknown.

It is possible that no catchpot was installed because the designers did not foresee that a runaway reaction could occur and that the relief device was intended to protect against other causes of overpressure such as overfilling or overpressurisation with the compressed air used to empty the reactor[8]. However, it is not good practice to discharge even relatively harmless liquids into the open air and there had been three runaway reactions on other similar plants which should have been known to the designers. One of the worst occurred at Ludwigshaven, Germany in 1953; 55 men were affected by dioxin, some seriously, and it was found to be impossible to decontaminate the building which had to be destroyed[7,9]. Another occurred in 1968 in Bolsover, Derbyshire, UK when the escaping gases exploded and one man was killed. In this case the reactor was heated by hot oil at 300°C and the manual heating control system failed[7].

Grossel[10] has reviewed the equipment available for containing discharges from relief systems.

The absence of a catchpot at Seveso nullified the other safety precautions as effectively as the lack of landward defences at Singapore nullified its seaward defences (see quotation at the head of this chapter). It 'was little more than an offshore island with a few guns pointing out to sea' (T. Hall[1]).

9.3   The official investigation

A year after Seveso the Italian Parliament set up a Parliamentary Commission of Inquiry and its report has been translated into English[11]. It criticised the company (ICMESA) for changing the procedure described in the original 1947 patent application; they had changed the ratio of the raw materials and the order in which various steps were carried out. However, according to Marshall[8], there is no evidence that this caused the runaway. The Commission, he says, did not seem to realise that all processes change with the passage of time and that the first description of a process is not sacrosanct.

The Italian procedure is thus very different from that followed in the UK where serious accidents such as Flixborough (see Chapters 8, 13 and 17–20) are investigated by a judge or senior lawyer, assisted by technical experts. Many other experts are asked to give evidence and the data are examined exhaustively.

9.4   Public response

The public response was similar to that of Flixborough, rather less in the UK, rather greater in the rest of Europe and it is not necessary to add anything to the discussion in Chapter 8.

Although dioxin is very toxic, one of the results of Seveso is that public concern over its hazards has greatly exceeded the real risk. The only confirmed deaths from dioxin poisoning occurred in the Netherlands in 1963 following a leak in a pharmaceutical plant. Of the fifty or so men

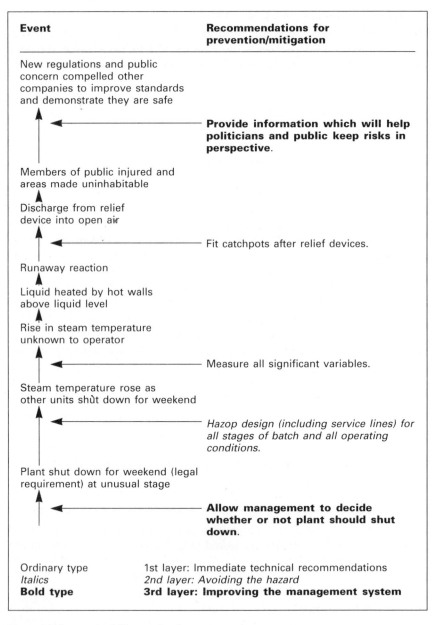

Event	Recommendations for prevention/mitigation
New regulations and public concern compelled other companies to improve standards and demonstrate they are safe	
	Provide information which will help politicians and public keep risks in perspective.
Members of public injured and areas made uninhabitable	
Discharge from relief device into open air	
	Fit catchpots after relief devices.
Runaway reaction	
Liquid heated by hot walls above liquid level	
Rise in steam temperature unknown to operator	
	Measure all significant variables.
Steam temperature rose as other units shut down for weekend	
	Hazop design (including service lines) for all stages of batch and all operating conditions.
Plant shut down for weekend (legal requirement) at unusual stage	
	Allow management to decide whether or not plant should shut down.
Ordinary type	1st layer: Immediate technical recommendations
Italics	*2nd layer: Avoiding the hazard*
Bold type	**3rd layer: Improving the management system**

Figure 9.2 Summary of Chapter 9 – Seveso

involved in cleaning up the spillage, estimated as 30–200 g, four died and about twelve developed chloracne[12]. This is an unpleasant disease but recovery is complete except for slight scarring.

Dioxin is formed by combustion and is very persistent. Traces are present in soil, in many manufactured products and in fish, meat and dairy products, the result of animals eating plankton or vegetation on which smoke particles containing dioxin have fallen. It is formed naturally by volcanos and forest fires and man-made sources include fires, furnaces and incinerators. Changes in incinerator design and operating conditions have greatly reduced the amount of dioxin they produce. There is no evidence that the minute traces present in paper and disposable nappies that have been bleached with chlorine are harmful[13].

9.5   Other similar incidents

Overheating of the bare metal, above a liquid layer, has caused a number of other incidents, though in a rather different way. The best known are Boiling Liquid Expanding Vapour Explosions or BLEVEs. These occur when a pressure vessel containing liquid is exposed to fire. The liquid cannot get above its boiling point and so, below the liquid level, the liquid keeps the walls of the vessel cool. However, if the flames impinge on the upper unwetted parts of the walls they soon become so hot that they soften and the vessel bursts even though the pressure is at or below the set point of the relief valve. If the liquid in the vessel is flammable the escaping vapour and liquid immediately ignite and produce an immense fireball[14].

A very different incident occurred in the reboiler of an ethylene oxide (EO) distillation column. The EO was in the tubes and the steam in the shell. Minor upsets during a start-up caused the liquid level in the tubes to fall. The temperature of the vapour above the liquid rose by about 60°C and approached that of the steam. A thin film of polymer peeled off the walls of the tubes and iron oxide in it catalysed a previously unknown reaction which made the vapour even hotter, so hot that the EO decomposed explosively. One man was killed, the distillation column was destroyed and the rest of the plant was damaged extensively[15].

Recommendations made after the explosion included actions to prevent a fall in the liquid level in the tubes and the use of the coolest possible heating medium.

Other incidents have occurred because relief devices discharged to atmosphere instead of a catchpot. For example, in Frankfurt, Germany in 1993 2 tonnes of chemicals were discharged through a relief valve. This prevented a vessel bursting but the chemicals damaged roofs and cars and contaminated soil and plants[16]. Relief valves are designed to lift so a discharge from one is a foreseeable event. If we were sure a relief valve would not lift we would not need to install it.

References

1 This quotation from the British Prime Minister on the fall of Singapore in 1942 is displayed in the museum on Sentosa Island, Singapore. T. Hall (*The Fall of Singapore*,

Methuen, Port Melbourne, Australia, 1983, p 71) writes, 'He had always believed what his generals had told him, that Singapore was impregnable. Now they said that what they meant was that it was impregnable providing the mainline was defended'.

2 European Community: Council Directive of 24 June 1982 on the Major Accident Hazards of Certain Industrial Activities, *Official Journal of the European Communities*, 1989, No. L230, 5 August 1982, pp. 1–18.

3 *Control of Industrial Major Accident Hazard Regulations*, Statutory Instrument No. 1902, Her Majesty's Stationery office, London, 1984. See also *A Guide to the Control of Industrial Major Accident Hazard Regulations* 1984, Her Majesty's Stationery Office, London, 1990.

4 Bretherick, L., *Loss Prevention Bulletin*, No. 054, Dec. 1983, p. 38.

5 Theofanous, T. G., *Nature*, Vol. 291, 25 June 1981, p. 640.

6 Cardillo, P. *et al.*, *Journal of Hazardous Materials*, Vol. 9, 1984, p. 221.

7 Bretherick, L., *Handbook of Reactive Chemical Hazards*, 4th edition, Butterworths, London, 1990, p. 562 (entry on 1,2,3,4–tetrachlorobenzene).

8 Marshall, V. C., *Loss Prevention Bulletin*, No. 104, April 1992, p. 15.

9 Lees, F. P., *Loss Prevention in the Process Industries*, Vol. 2, Butterworths, London, 1980, Appendix 2.

10 Grossel, S. S., *Journal of Loss Prevention in the Process Industries*, Vol 3, No. 1, Jan. 1990, p. 112.

11 *Translation of the Italian Parliamentary Commission into the Seveso Disaster*, Health and Safety Executive, Sheffield, UK.

12 Withers, J., *Major Industrial Hazards*, Gower, Aldershot, UK, 1988, p. 34.

13 *Environmental Issues*, Imperial Chemical Industries, Runcorn, UK, Dec. 1991.

14 Kletz, T. A., *What Went Wrong – Case Histories of Process Plant Disasters*, 2nd edition Gulf, Houston, Texas, 1988, Section 8.1.

15 Viera, G. A. and Wadia, P. H., 'Ethylene oxide explosion at Seadrift, Texas. Part 1 – Background and technical findings', *Proceedings of the American Institute of Chemical Engineers 27th Loss Prevention Symposium*, March 1993.

16 *The Chemical Engineer*, No. 538, 11 March 1993, p. 8.

Chapter 10

Bhopal

The gas leak just can't be from my plant. The plant is shut down. Our technology just can't go wrong, we just can't have such leaks.'
                    The first reaction of the Bhopal works manager[1].

The worst accident in the history of the chemical industry occurred in the Union Carbide plant at Bhopal, India on 3 December 1984 when a leak of over 25 tonnes of methyl isocyanate (MIC) from a storage tank spread beyond the plant boundary, killing over 2000 people. The official figure was 2153[2] but according to some reports the true figure was nearer 10 000[3,4]. In addition about 200 000 people were injured. Most of those killed and injured were living in a shanty town that had grown up close to the plant.

Before 1984 the worst accidents that had occurred in the chemical industry were the explosion of a 50/50 mixture of ammonium sulphate and ammonium nitrate at Oppau in Germany in 1921, which killed 430 people including fifty members of the public, and the explosion of a cargo of ammonium nitrate in a ship in Texas City harbour in 1947, which killed 552 people, mostly members of the public[5]. If conventional explosives are classified as chemicals then we should include the explosion of an ammunition ship in Halifax, Novia Scotia in 1917 which killed about 1800 people. Earlier in 1984, on 19 November, 550 people were killed when a fire occurred at a liquefied petroleum gas processing plant and distribution centre in San Juanico, a suburb of Mexico City[6] and on 25 February at least 508 people, most of them children, were killed when a petrol pipe ruptured in Cubatao, near Sao Paulo, Brazil and the petrol spread across a swamp and caught fire[7]. In both cases most of those killed were living in shanty towns. 1984 was thus a bad year for the chemical industry.

The Bhopal tragedy started when a tank of MIC – an intermediate used in the manufacture of the insecticide carbaryl – became contaminated with water – probably as the result of sabotage – and a runaway reaction occurred. The temperature and pressure rose, the relief valve lifted and MIC vapour was discharged into the atmosphere. The protective equipment which should have prevented or minimised the discharge was out of action or not in full working order: the refrigeration system which should have cooled the storage tank was shut down, the scrubbing system which should have absorbed the vapour was not immediately available and the flare system, which should have burnt any vapour which got past the scrubbing system, was out of use.

10.1 'What you don't have, can't leak'

There are many lessons to be learnt from Bhopal but the most important is that the material which leaked need not have been there at all. It was an intermediate, not a product or raw material, and while it was convenient to store it, it was not essential to do so. Originally MIC was imported and had to be stored but later it was manufactured on site. Nevertheless over 100 tonnes were in store, some of it in drums. After the tragedy it was reported that Du Pont intended to eliminate intermediate storage from a similar plant that they operated. Instead they use the MIC as soon as it is produced, so that instead of 40 tonnes in a tank there will be only 5–10 kg in a pipeline[8]. Mitsubishi were said to do this already[8,9]. During the next few years other companies described plans for reducing their stocks of other hazardous intermediates[10]. At the end of 1985 Union Carbide claimed that the inventories of thirty-six toxic chemicals had been reduced by 74%[11].

The main lesson of Bhopal is thus the same as that of Flixborough: 'What you don't have, can't leak'. Whenever possible, we should reduce or eliminate inventories of hazardous materials, in process or storage. It is unfortunate, to say the least, that more notice was not taken of the papers written after Flixborough which stressed the desirability of inherently safer designs, as they are called[12-14]. It seems that most companies felt so confident of their ability to keep hazardous materials under control that they did not look for ways of reducing inventories. Yet to keep lions under control we need expensive added-on protective equipment which may fail or may be neglected. It is better to keep lambs instead.

Intermediate storage is usually wanted by plant managers as it makes operation easier: one section of the plant can continue on line when another is shut down. Computer studies on equipment availability always show that intermediate storage is needed. However, they do not allow for the fact that if intermediate storage is available it will always be used, and maintenance will be carried out at leisure, but if it is not available people do everything possible to keep the plant on line, carrying out maintenance as soon as possible. The need for intermediate storage is thus a self-fulfilling prophecy.

Another alternative to intermediate storage is to build a slightly larger plant and accept fewer running hours per year. Intermediate storage, including working capital, is expensive, as well as hazardous, but this alternative is rarely considered.

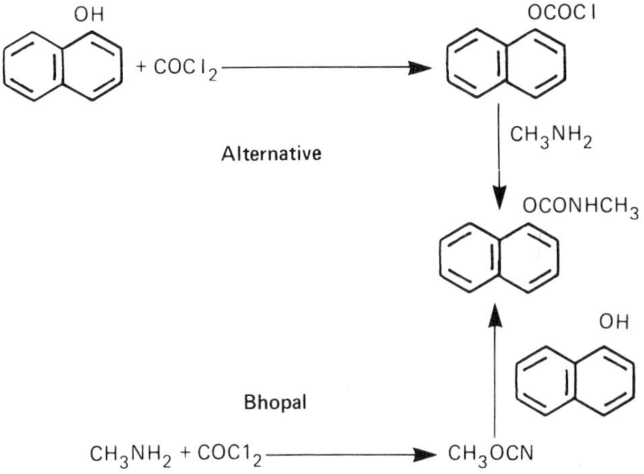

Figure 10.1 Routes to carbaryl

If reducing inventories, or intensification as it is called, is not practicable an alternative is substitution, that is, using a safer material or route. At Bhopal the product (carbaryl) was made from phosgene, methylamine and alpha-naphthol. The first two were reacted together to make MIC which was then reacted with alpha-naphthol. In an alternative process used by the Israeli company Makhteshim, alpha-naphthol and phosgene are reacted together to make a chloroformate ester which is then reacted with methylamine to make carbaryl. The same raw materials are used but MIC is not formed at all (Figure 10.1)[15].

Of course, phosgene is also a hazardous material and its inventory should be kept as low as possible, or avoided altogether. If carbaryl can only be made via phosgene, perhaps another insecticide should be manufactured instead.

Or instead of manufacturing pesticides perhaps we should achieve our objective – preventing the harm done by pests – in other ways, for example, by breeding pest-resistant plants or by introducing natural predators? I am not saying we should, both these suggestions have disadvantages, merely saying that the question should be asked.

10.2   Plant location

If materials which are not there cannot leak, people who are not there cannot be killed. The death toll at Bhopal would have been much smaller if a shanty town had not been allowed to grow up near the plant. In many countries planning controls prevent such developments, but not apparently in India. Of course, it is much more difficult to prevent the growth of shanty towns than of permanent dwellings, but nevertheless it is essential to stop them springing up close to hazardous plants. If the government or local authorities cannot control their growth then industry should be prepared, if necessary, to buy up the land and fence it off. As already mentioned the high death tolls at Mexico City and Cubatao earlier in 1984 occurred in shanty towns.

It is generally agreed that if a leak of toxic gas occurs people living in the path of the plume should stay indoors with their windows closed. Only if the leak is likely to continue for a long time, say, more than half-an-hour, should evacuation be considered. However, this hardly applies to shanties which are usually so well ventilated that the gas will enter them at once. It may be more difficult to prevent the growth of shanty towns than of permanent buildings, but it is also more important to do so.

The need to keep hazardous plants and concentrations of people apart has been known for a long time. In the 19th century a series of explosions in fireworks and explosives factories in the UK led ultimately to regulations on location and an inspectorate to enforce them. In 1915, however, the government insisted that TNT should be manufactured in a factory at Silvertown, a heavily built-up area near London. The owners of the factory, Brunner Mond, objected but were overruled as explosives were urgently needed to support the war effort. In 1917 54 tons of TNT exploded, devastating the site and the surrounding area. Seventy-three people were killed, including everyone working in the factory, and a hundred were seriously injured. Eighty-three houses were flattened or so badly damaged that they had to be demolished and 765 were seriously damaged and needed new interiors. After the explosion the officials who had insisted that production of TNT at Silvertown was essential now said that the loss of the factory would make no practical difference to the output of munitions[16].

The immediate cause of the explosion is unknown but the underlying cause was the erroneous belief that TNT was not very dangerous. It was not covered by the Explosives Act, although another TNT factory had blown up in 1915. Compare the quotation at the head of this chapter.

10.3 Why did a runway reaction occur?

The MIC storage tank became contaminated by substantial quantities of water and chloroform, up to a tonne of water and 1½ tonnes of chloroform, and this led to a complex series of runaway reactions, a rise in temperature and pressure and discharge of MIC vapour from the storage tank relief valves[17]. The precise route by which the water entered the tank is not known, though several theories have been put forward[18,19]. One theory is that it came from a section of vent line some distance away that was being washed out. This vent line should have been isolated by a slip-plate (spade) but it had not been fitted. However, the water would have had to pass through six valves in series and it seems unlikely that a tonne could have entered the tank in this way. Another theory is that the water entered via the nitrogen supply line. Kalelkar has argued convincingly that there was a deliberate act of sabotage by someone who did not realise what the results of his actions would be[20]. This theory has been criticised by many people who sympathised with the victims and thought an attempt was being made to whitewash Union Carbide. This was not the case. The company is still responsible for stocking more MIC than is necessary, for allowing a shanty town to grow alongside the plant, for not keeping protective equipment in working order (see next section) and for all the other deficiencies discussed later in this chapter.

When we are looking for the underlying causes of the accident, rather than the immediate cause, the route by which the water entered the MIC tank hardly matters. Since it is well-known that water reacts violently with MIC, no water should have been allowed anywhere near the equipment, for washing out lines or for any other purpose. If water is not there, it cannot leak in or be added deliberately. If any of the suggested routes were possible, then they should have been closed before the disaster occurred. So far as is known, no hazard and operability study (hazop)[21,22] was carried out on the design though hazop is a powerful tool, used by many companies for many years, for identifying routes by which contamination (and other unwanted deviations) can occur.

10.4   Keep protective equipment in working order – and size it correctly

The storage tank was fitted with a refrigeration system but it was not in use. The scrubbing system which should have absorbed the MIC discharged through the relief valve was not in full working order. The flare system which should have burned any MIC which got past the scrubbing system was disconnected from the plant for repair. The high temperature and pressure on the MIC tank were at first ignored, as the instruments were poorly maintained and known to be unreliable[23]. The high temperature alarm did not operate as the set point had been altered and was too high[24]. One of the main lessons from Bhopal is thus the need to keep all protective equipment in full working order.

It is easy to buy safety equipment; all we need is money and if we make enough fuss we get it in the end. It is much more difficult to make sure that the equipment is kept in full working order, especially when the initial enthusiasm has worn off. Procedures, including testing and maintenance procedures, are subject to a form of corrosion more rapid than that which affects the steelwork and can vanish without trace in a few months once managers lose interest. A continuous auditing effort is needed by managers at all levels to make sure that procedures are maintained (see Chapter 6).

Sometimes procedures lapse because managers lose interest. Unknown to them, operators discontinue safety measures. At Bhopal it went further than this. Disconnecting the flare system and shutting down the refrigeration system are hardly decisions that operators are likely to take on their own. The managers themselves must have taken these decisions and thus shown a lack of understanding and/or commitment.

It is possible that the protective equipment was out of use because the plant that produced the MIC was shut down and everyone assumed that the equipment had been installed to protect the plant rather than the storage. Runaway reactions, leaks, and discharges from relief valves are commoner on plants than on storage systems but they are by no means unknown on storage systems. Twenty-four of the hundred largest insurance losses in a thirty-year period occurred in storage areas and their value was higher than average[25]. Furthermore, since a relief valve was installed on the storage tank, it was liable to lift and the protective equipment should have been available to handle the discharge. As stated in Chapter 9, if the designers were sure that a relief valve would never lift there would have been no need to install it.

It has been argued that the refrigeration, scrubbing and flare systems were not designed to cope with a runaway reaction of the size that occurred and that there would have been a substantial discharge of MIC to atmosphere even if they had all been in full working order. This may be so, but they would certainly have reduced the size of the discharge and delayed its start.

The relief valve was too small for the discharge from a runaway reaction. The pressure in the storage vessel, designed for a gauge pressure of 40 psi (2.7 bar), reached 200–250 psi (14–17 bar). The vessel was distorted and nearly burst. If it had burst the loss of life might have been lower as there would have been less dispersion of the vapour. The relief valve was designed to handle vapour only but the actual flow was a two-phase mixture of vapour and liquid[26].

If the protective equipment was not designed to handle a runaway, or two-phase flow, we are entitled to ask why. Were the possibilities of a runaway or two-phase flow not foreseen or were they considered so unlikely that it was not necessary to guard against them? What formal procedures were used during design to answer these questions?

Although the managers (and also the operators, but they take their cue from the managers) showed less competence and commitment to safety than might reasonably have been expected, we should not assume that Indian managers (and operators) are in general less competent than those in the west. There are poor managers in every country and there is no reason to believe than the standard in India is any lower than elsewhere. In one respect the managing director of Union Carbide India showed more awareness than his US colleagues: he queried the need for so much MIC storage but was overruled[18].

Bhopal illustrates the limitations of hazard assessment techniques. If asked, before the accident, to estimate the probability of a leak of MIC, by fault tree or similar techniques, most analysts would have estimated the failure rates of the refrigeration, scrubbing and flare systems but would not have considered the possibility that they might all be switched off. Hazard assessments become garbage if the assumptions on which they are based are no longer true.

Similarly, estimates of human error are usually estimates of the probability that a man will forget to carry out a task, such as closing a valve, or carry it out wrongly. We cannot estimate the probability that he will make a conscious decision not to close it, either because he considers it unnecessary to do so or because he wishes to sabotage operations[27].

Some reports on Bhopal suggested that the instrumentation there was less sophisticated than on similar plants in the United States and that this may have led to the accident. This is a red herring. If conventional instrumentation was not adequately maintained and its warnings were ignored, then there is no reason to believe that computerised instrumentation would have been treated any differently. In fact the reverse may be the case. If people are unable or unwilling to maintain basic equipment, they are less likely to maintain sophisticated equipment. Nevertheless, during the investigation of accidents which have occurred because the safety equipment provided was not used, people often suggest that more equipment is installed (see Chapter 6).

Another protective device was a water spray system which was designed to absorb small leaks at or near ground level. It was not intended to absorb relief valve discharges at a high level and failed to do so.

10.5   Joint ventures

The Bhopal plant was half-owned by a US company, Union Carbide, and half-owned locally. Although Union Carbide had designed the plant and had been involved in the start up, by the time of the accident the Indian company had become responsible for operations, as required by Indian law.

In the case of such joint ventures it is important to be clear as to who is responsible for safety in design and operation. The technically more sophisticated partner has a special responsibility if it is not directly responsible. It should make sure that the operating partner has the knowledge, skill, commitment and resources necessary for safe operation. If not, it should not go ahead with the venture. It cannot shrug off responsibility by saying that it is no longer in full control. Soon after Bhopal one commentator wrote, '... multinational companies and their host countries have got themselves into a situation in which neither feels fully responsible'[28].

People who sell or give dangerous articles to children are responsible if the children injure themselves or others. Similarly, if we give dangerous plant or material to people who have not demonstrated their competence to handle it we are responsible for the injuries they cause.

For Union Carbide the Bhopal plant was a backwater, making little contribution to profits, in fact often losing money, and may have received less than its fair share of management resources[29].

At Flixborough (Chapter 8) the partner with knowledge of the technology (Dutch State Mines) was in control.

10.6   Training in loss prevention

Bhopal makes us ask if the people in charge of the plant, and those who designed it, received sufficient training in loss prevention, as students and from their employers. In the United Kingdom all chemical engineering undergraduates receive some training in loss prevention[30]. If they do not they are not able to join the Institution of Chemical Engineers as full or corporate members. In most other countries, including the United States, most undergraduate chemical engineers receive no such training, although the American Institute of Chemical Engineers is now encouraging universities to introduce it.

There are several reasons why loss prevention should be included in the training of chemical engineers[31]:

(1) Loss prevention should not be something added on to a plant after design like a coat of paint but an integral part of design. Hazards should, whenever possible, be removed by a change in design, such as reduction in inventory, rather than by adding on protective equipment. The designer should not ask the safety adviser to add on the safety features for him; he should be taught to design a plant which does not require added-on safety features.
(2) Most engineers never use much of the knowledge they acquire as students but almost all have at some time to take decisions on loss prevention. Universities which give no training in loss prevention are not preparing their students for the tasks they will have to undertake.
(3) Loss prevention can be used to illustrate many of the principles of chemical engineering and to show that many problems which at first sight do not seem to lend themselves to numerical treatment can in fact be treated quantitatively.

Since in many countries universities are not providing training in chemical engineering, companies should make up the deficiency by internal training. Many try to but often in rather a haphazard way – an occasional course or lecture. Few companies put all new recruits through a planned programme.

At Bhopal the original managers had left and had been replaced by others whose experience had been mainly in the manufacture of batteries. There had been eight different managers in charge of the plant in 15 years[32]. Many of the original operators had also left and one wonders how well their successors were trained[33].

However, while these facts, and reductions in manning, may be evidence of poor management and a lack of commitment to safety I do not think that they contributed directly to the accident. The errors that were made, such as disconnection of safety equipment and resetting trips at too high a level, were basic ones that cannot be blamed on inexperience of the particular plant. No manager who knew and accepted the first principles of loss prevention would have allowed them to occur.

10.7   Handling emergencies

More than any other accident described in this book Bhopal showed up deficiencies in the procedures for handling emergencies, both by the company and by the local authorities. It showed clearly the need for companies to collaborate with the emergency services in identifying incidents that might occur and their consequences, drawing up plans to cope with them and exercising these plans. In the United Kingdom this is now required by law[34]. This aspect is discussed in reference 35.

10.8   Public response

In the United States Bhopal produced a public reaction similar to that produced in the United Kingdom by Flixborough (Chapter 8) and in the rest of Europe by Seveso (Chapter 9). Many companies spent a great deal of money and effort making sure that a similar accident could not occur on their plants. Even so, Bhopal seems to have produced less 'regulatory fallout' than Flixborough or Seveso. The US chemical industry has tried to convince the authorities that it can put its own house in order. In particular the American Institute of Chemical Engineers set up a Center for Chemical Process Safety, generously funded by the chemical industry, to provide advice on loss prevention. One of its objectives is to have loss prevention included in the training of undergraduates. The Chemical Manufacturers Association has launched a Community Awareness and Response (CAER) programme to encourage companies to improve their emergency plans and a National Chemical Response Information Center to provide the public and emergency services with advice and assistance before and during emergencies.

Nevertheless in a paper called 'A field day for the legislators' Stover[36] lists thirty-two US Government proposals or activities and thirty-five international activities that had been initiated by the end of 1985. In addition there have been state and local responses in the US. These have ben reviewed by Horner[37].

In India there were, of course, extensive social effects. They are reviewed in reference 38.

After Bhopal many international companies reviewed the extent to which they controlled and audited their subsidiaries. Chapters 4 and 5 also

Events	Recommendations for prevention/mitigation

Public concern compelled other companies to improve standards

← **Provide information that will help public keep risks in perspective.**

Emergency not handled well

← Provide and practise emergency plans.

Over 2000 people killed

← **Control building near major hazards.**

Scrubber not in full working order
Flare stack out of use
Both may have been too small

← Keep protective equipment in working order. Size for foreseeable conditions.

Discharge from relief valve

Refrigeration system out of use

← Keep protective equipment in use even though plant is shutdown.

Runaway reaction

Rise in temperature

← **Train operators not to ignore unusual readings.**
Keep instruments in working order.

Water entered MIC tank

← *Carry out hazops on new designs.*
Do not allow water near MIC.

Decision to store over 100 t MIC

← *Minimise stocks of hazardous materials.*

Decision to use MIC route

← *Avoid use of hazardous intermediates.*

Decision to manufacture carbaryl

← **Consider less hazardous alternatives.**

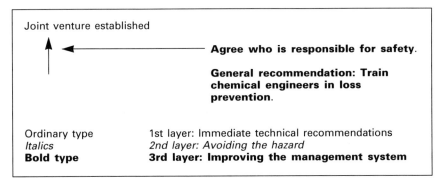

Figure 10.2 Summary of Chapter 10 – Bhopal

showed what can happen when they are left to do as they wish. Whiston has described the changes made by ICI in the years after Bhopal[39].

Terrible though Bhopal was, we should beware of over-reaction and of suggestions that insecticides, or even the whole chemical industry, are unnecessary. Insecticides, by increasing food production, have saved far more people than Bhopal has killed. But Bhopal was not an inevitable result of insecticide manufacture. By better design or by better operation, by just one of the recommendations summarised in Figure 10.2, Bhopal could have been prevented. The most effective methods of prevention are those near the bottom of the diagram, such as reduction in inventory or change in the process. The safety measures at Bhopal, such as the scrubber and the flare stack, were too near the top of the chain, too near the top event. If they failed there was nothing to fall back on. To prevent the next Bhopal we need to start at the bottom of the chain.

References

1 Bidwai, P., quoted by Bogard, W., *The Bhopal Tragedy*, Westview Press, Boulder, Colorado, 1989, p. 14.
2 Tachakra, S. S., *Journal of Loss Prevention in the Process Industries*, Vol. 1, No. 1, January 1988, p. 3.
3 *Free Labour World*, International Federation of Free Trade Unions, Brussels, Belgium, No. 1/86, 18 January 1986, p. 1.
4 Shrivastava, P., *Bhopal – Anatomy of a Crisis*, Ballinger, Cambridge, Massachusetts, 1987, p. 64.
5 Lees, F. P., *Loss Prevention in the Process Industries*, Vol. 2, Butterworths, London, 1980, Appendix 3.
6 *BLEVE – The Tragedy of San Juanico*, Skandia International, Stockholm, 1985.
7 *Hazardous Cargo Bulletin*, June 1984, p. 34.
8 *Chemical Week*, 23 January 1985, p. 8.
9 *Chemistry in Britain*, Vol. 21, No. 2, Feb. 1985, p. 123.
10 Wade, D. E., *Proceedings of the International Symposium on Preventing Major Chemical Accidents*, American Institute of Chemical Engineers, Washington, DC, 3–5 February 1987, Paper 2.1.

11 *Chemical Insight*, Late Nov. 1985, p. 1.

12 Kletz, T. A., *Chemical Engineering*, Vol. 83, No. 8, 12 April 1976, p. 124.

13 Kletz, T. A., *Chemistry and Industry*, 6 May 1978, p. 37.

14 Kletz, T. A., *Hydrocarbon Processing*, Vol. 59, No. 8, Aug. 1980, p. 137.

15 Reuben, B. M., Private communication.

16 Neal, W., *With Disastrous Consequences. . .*, Hisarlik Press, London, 1992, Chapters 3 and 7.

17 *Bhopal Methyl Isocyanate Incident: Investigation Team Report*, Union Carbide Corporation, Charleston, South Carolina, March 1985.

18 *The Trade Union Report on Bhopal*, International Federation of Free Trade Unions and International Federation of Chemical, Energy and General Workers' Unions, Geneva, Switzerland, 1985.

19 Varadarajan, S. *et al.*, *Report on Scientific Studies on the Factors related to Bhopal Toxic Gas Leakage*, Indian Planning Commission, Delhi, India, Dec. 1985.

20 Kalelkar, A. S., 'Investigations of large magnitude incidents – Bhopal as a case study', *Preventing Major Chemical and Related \Process Accidents*, Symposium, Series No 110, Institution of Chemical Engineers, Rugby, UK, 1988, p. 553.

21 Kletz, T. A., *Hazop and Hazan – Identifying and Assessing Process Industry Hazards*, 3rd edition, Institution of Chemical Engineers, Rugby, UK, 1992.

22 Knowlton, R. E., *A Manual of Hazard and Operability Studies*, Chemetics International, Vancouver, Canada, 1992.

23 *New York Times*, 28 Jan–3 Feb 1985.

24 Shrivastava, P., *The Accident at Union Carbide Plant in Bhopal – A Case Study*, Air Pollution Control Association Conference on Avoiding and Managing Environmental Damage from Major Industrial Accidents, Vancouver, Canada, 3–6 Nov. 1985.

25 *One Hundred Largest Losses*, Marsh and McLennan, Chicago, Illinois, 8th edition, 1985.

26 Swift, I., in *The Chemical Industry after Bhopal, Proceedings of a Symposium*, London, 7/8 November 1985, IBC Technical Services.

27 Kletz, T. A., *An Engineer's View of Human Error*, 2nd edition, Institution of Chemical Engineers, Rugby, UK, 1991, Chapter 5.

28 Smith, A. W., *The Daily Telegraph*, 15 Dec. 1984, p. 15.

29 As ref. 4, p. 51.

30 *First Degree Course including Guidelines on Accrediting of Degree Course*, Institution of Chemical Engineers, Rugby. UK, 1989.

31 Kletz, T. A., *Plant/Operations Progress*, Vol. 7, No. 2, April 1988, p. 95.

32 As ref. 4, p. 52.

33 *The Bhopal Papers*, Transnationals Information Centre, London, 1986, p. 4.

34 *Control of Industrial Major Accident Hazard Regulations*, Statutory Instrument No. 1902, Her Majesty's Stationery Office, London, 1984.

35 As ref. 4, Chapter 6.

36 Stover, W., in *The Chemical Industry after Bhopal, Proceedings of a Symposium*, London, 7/8 November 1985, IBC Technical Services,

37 Horner, R. A., *Journal of Loss Prevention in the Process Industries*, Vol. 2, No. 3, July 1989, p. 123.

38 As ref. 4, Chapters 4 and 5.

39 Whiston, J., 'Corporate SHE policy and implementation of international SHE standards', *Health, Safety and Loss Prevention in the Oil, Chemical and Process Industries*, Butterworth-Heinemann, Oxford, 1993, p. 19.

Chapter 11

Three Mile Island

...more disasters are caused by those who command than by those who fortuitously turn the wrong knob or dump the wrong chemical.

P. T. W. Hudson[1]

This chapter and the next one describe two nuclear accidents that have many lessons for other industries. Recommendations of purely nuclear interest are not discussed.

On 28 March 1979 the nuclear power station at Three Mile Island in Pennsylvania overheated and a small amount of radioactivity escaped to the atmosphere. Even in the long term no one is likely to be harmed but nevertheless the incident shattered public confidence in the nuclear industry and led to widespread demands for a halt to its expansion.

Figure 11.1 is a simplified diagram of a pressurised water reactor (PWR), the type used at Three Mile Island and in most nuclear power stations. Existing UK reactors are gas-cooled but a PWR is being built at Sizewell. Heat is generated in the core by nuclear fission and is removed by pumping primary water round and round. The water is kept under pressure so that it is not boiling. (It is called a pressurised water reactor to distinguish it from a boiling water reactor, the type used at Chernobyl.) The primary water gives up its heat to the secondary water which does boil. The steam produced drives a turbine and is condensed. The condensate is recycled. All radioactive materials, including the primary water, are enclosed in a containment building so that they will not escape if there is a leak[2-10].

11.1 Phase A – The trouble starts

The secondary water passed through a resin polisher unit to remove traces of impurities. There were several parallel paths and one of them choked. Less attention had been paid to the design of this off-the-shelf ancillary unit than to the design of the main radioactive equipment. In particular its reliability had not been studied to the same extent.

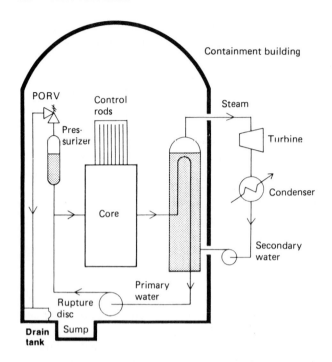

Figure 11.1 A pressurized-water reactor – simplified

The first lesson to be learned, therefore, is that *packaged units, ancillary units, off-plots and so on need as much attention as the main stream,* especially when their failure can cause a shutdown of the main stream.

To try to clear the choke the operators used instrument air. Its pressure was lower than that of the water so some water got back into the instrument air lines.

The second lesson is that we should *never connect service lines to process equipment at a higher pressure.* This has often been done and another case is described in Sections 2.1 and 2.4, which also list the precautions that should be taken if the pressures of the process or services lines are liable to change. In addition *it is bad practice to use instrument air lines for line blowing.* Ordinary compressed air should be used instead as the results of any contamination of the air are then less serious.

There was a non-return valve in the compressed air line but it was faulty. Non-return valves have a poor reputation but in many plants they are never inspected or maintained. We cannot expect any piece of equipment, especially one containing moving parts, to operate correctly for ever without attention. On non-nuclear plants *non-return valves should be scheduled for regular inspection,* say every year or two. However non-return valves are designed to prevent gross flow, and should not be relied on to prevent a slight flow.

Fluidic non-return valves, which contain no moving parts, should be considered in place of conventional ones[11]. However, their normal leak rate is higher.

The water in the instrument air lines caused several instruments to fail and the turbine tripped. This stopped heat removal from the radioactive core. The production of heat by fission stopped automatically within a few minutes. (Metal rods dropped down into the core. They absorb neutrons and stop radioactive fission.) However, some heat was still produced by radioactive decay, at first about 6% of the total load, falling to about 1.5% after an hour, and this caused the primary water to boil. The pilot-operated relief valve (PORV) on the primary circuit lifted and pumps started up automatically to replace the water evaporated from the primary circuit.

Unfortunately the PORV stuck open.

11.2   Phase B – Things worsen

The operators did not realise that the PORV was stuck open because a light on the panel told them it was shut. However, the light was not operated by the valve position but by the signal to the valve. The operators had not been told this or had forgotten.

Whenever possible *instruments should measure directly what we want to know*, not some other property from which it can be inferred. If a direct measurement is impossible, then the label on the panel should tell us what is measured, in this case, 'Signal to PORV', not 'PORV position' (see Section 11.8).

Several other readings should have suggested to the operators that the PORV was stuck open and that the water in the primary circuit was boiling:

(1) The PORV exit line was hotter than usual, 140°C instead of 90°C, but this was thought to be due to residual heat.
(2) The pressure and temperature of the primary water were lower than usual.
(3) There was a high level in the containment building sump.
(4) The primary water circulation pumps were vibrating.

On the other hand the level in the pressuriser was high as it was raised by bubbles of steam.

The operators decided to believe the PORV position light and the pressuriser level and ignore or explain away the other readings, probably because (a) they did not really understand how the temperature and pressure in the primary circuit depended on each other and when boiling would occur, and (b) their instructions and training had emphasised that it was dangerous to allow the primary circuit to get too full of water. They had not been told what to do if there was a small leak of primary water (although they had been told what to do if there was a major leak such as a pipe fracture).

The operators thought the PORV was shut. Conditions were clearly wrong and their training had emphasised that too much water was dangerous. They therefore shut down the water pumps. Note that the only action taken by the operators made matters worse. If they had done nothing, the plant would have cooled down safely on its own.

However, it is superficial to say that the accident was due to operator error. The operators were working under stress and were inadequately trained; a hundred alarms were indicating an alarm condition and the crucial instrument was misleading. The work situation – that is, the design and the management system – were at fault, not the operators. To quote Reason[12]:

> Rather than being the main instigators of an accident, operators tend to be the inheritors of 'pathogens' created by poor design, incorrect installation, faulty maintenance, inadequate procedures, bad management decisions, and the like. The operators' part is usually that of adding the final garnish to a lethal brew that has been long in the cooking. In short: unsafe acts in the front-line stem in large measure from bad decisions made by the rear echelons.

People sometimes say that automatic response is necessary if action has to be taken quickly but that we can rely on operators if they have plenty of time, say half-an-hour or more. However, this is true only if the operators can diagnose what is happening and know what action to take.

Three Mile Island shows us that we cannot operate complex plant by writing a series of precise instructions which must be followed to the letter. Problems will arise which are not foreseen or instruments will give conflicting results. Operators therefore need:

- To understand what goes on in the plant.
- To be trained in diagnostic skills.
- To be be given diagnostic aids.

One method of diagnostic training has been described by Duncan and co-workers[13,14]. The operator is shown a copy of the control panel on which certain readings are marked. He is asked to diagnose the fault and say what action he would take. The problems gradually increase in difficulty. The operator learns to identify and correct foreseeable faults and also learns general diagnostic skills that enable him to identify faults which have not been foreseen. Such training could have prevented Three Mile Island. (The fault which occurred there had not been foreseen though it should have been.)

Expert systems can be used to aid diagnosis[15].

Three Mile Island was not the first plant at which a PORV had stuck open. Similar incidents had happened before though with less serious consequences. But unfortunately the lessons of these incidents had not been passed on. We pay a high price for the information that accidents give us – deaths, injuries and damage to plant. We should make full use of that knowledge[16].

11.3   Phase C – Damage

With the make-up water isolated the level in the primary circuit fell. The top of the radioactive core was uncovered, steam reacted with the zirconium alloy cans which protect the uranium and hydrogen was formed. At the same time the steam which was being discharged through the PORV condensed in a drain tank, overflowed into the containment building sump and was automatically pumped outside the containment building.

Changes in design to minimise these consequences were recommended but they are not of general interest.

Damage did not occur until two hours after the start of the incident. The operators could have prevented the damage if, at any time during this period they had realised that the PORV was open and that the water level was low. However, they had made their diagnosis and they stuck to it, even though the evidence against it was overwhelming. They had developed a mind-set.

This happens to us all. We have a problem. We think of an explanation, sometime the first explanation that comes to mind. We are then so pleased with ourselves for solving the problem that we fail to see that our explanation is not fully satisfactory. Reference 17 describes some other accidents that occurred as the result of mind-sets. They are very difficult to avoid. Perhaps it would help if our training included examples of mind-sets so that we are aware that they occur and are thus a little less ready to grab the first explanation that we think of.

11.4   Following the rules is not enough

As already stated, it was believed at Three Mile Island, to a large extent, that all the operators had to do was to follow the rules. There was something of the same attitude amongst the managers and designers. They seem to have believed that if they followed the rules laid down by the US Nuclear Regulatory Commission they would have a safe plant. There is much less of this attitude in UK industry where instead of a lot of detailed rules and regulations there is a general requirement, under the Health and Safety at Work Act, to provide a safe plant and system of work and adequate instruction, training and supervision, so far as is reasonably practicable. It is the responsibility of those who manage the plant, and know most about it, to say what is a safe plant, etc., but if he does not agree the factory inspector will say so and, if necessary, issue an improvement or prohibition notice. If there is a generally accepted code of practice this should be followed, unless the employer can show that it is inapplicable or he is doing something else which is as safe or safer. Nevertheless, arguments in favour of more regulation surface from time to time in the UK so it may be worthwhile summarising the case for the UK approach:

- Codes are more flexible than regulations. They can be changed more easily when new problems arise or new solutions are found to old problems.
- We do not have to follow to the letter regulations which are inappropriate or out-of-date.
- Managers (and factory inspectors) cannot hide behind the regulations. They cannot say, 'My plant must be safe because I have followed all the rules'.
- It is not possible to write regulations to cover the details of complex and rapidly-changing technologies.
- The factory inspector has more, not less, power under the UK system. He does not have to prove that a regulation has been broken. It is sufficient for him to say that the plant or method of working is not safe. His case is stronger if a generally accepted code of practice is not being followed.

In brief, *designers should have to demonstrate that their designs are safe, not just follow the rules*.

11.5   Consider minor failures as well as major ones

At Three Mile Island there was much concern with major failures, such as complete fracture of a primary water pipe, but smaller and more probable accidents were ignored. There was a belief that if large-scale accidents can be controlled, minor ones can be controlled as well. This is not true, and as result what started as a minor incident became a major one. Similarly in the process industries most injuries and damages are caused by minor failures, not major ones, and failures of both sorts should be considered (see Chapter 15).

11.6   Public response

Three Mile Island produced a greater public reaction than any other incident described in this book, with exception of Bhopal and Chernobyl. The US nuclear industry received a setback from which it has not yet recovered and in other countries the accident has been extensively quoted to show that the nuclear industry is not as competent as it had led us to believe and that improbable accidents can happen; better therefore to stick to the hazards we know, such as coal mining. One of the lessons of Three Mile Island is that emergency plans should include plans for briefing the press and providing simple explanations of what has occurred and the extent of the risk.

Although no one was killed as a direct result of Three Mile Island nor by the subsequent radioactive fallout, because the plant was shut down more coal has had to be mined and burnt and this will have caused a few extra deaths by accident and pollution, perhaps two per year.

Three Mile Island was a new reactor and fission products had not had time to build up to the equilibrium level. If the reactor had been older the release of radioactivity would have been greater[18].

11.7   Must nuclear power stations be so hazardous?

I have left until near the end the most important lesson to be learned from Three Mile Island, one ignored by most commentators. The lesson is the same as that taught us by Bhopal (Chapter 10): whenever possible we should design plants that are inherently safer rather than plants which are made safe by adding on protective equipment which may fail or may be neglected. At Bhopal instead of storing the intermediate MIC, it should have been used as soon as it was produced, or the product should have been manufactured by a route that did not involve MIC. In designing nuclear power stations we should prefer those designs which are least dependent on added-on engineered safety systems to prevent overheating of the core and damage to it if coolant flow is lost. The damage at Three Mile Island was, of course, much less serious than that which might have occurred or that which occurred at Chernobyl (Chapter 12).

Gas-cooled reactors are inherently safer than water-cooled ones as if the coolant pressure is lost convection cooling, assisted by a large mass of graphite, prevents serious overheating. In addition, the operators have more time in which to act and are therefore less likely to make mistakes. Franklin writes, 'when operators are subject to conditions of extreme urgency ... they will react in ways that lead to a high risk of promoting accidents rather than diminishing them. This is materially increased if operators are aware of the very small time margins that are available to them' and '... It is much better to have reactors which, even if they do not secure the last few percent of capital cost effectiveness, provide the operator with half-an-hour to reflect on the consequences of the action before he needs to intervene'[19]. (At Three Mile Island the operators did have time, before damage occurred, to reconsider their decision to shut down the water pumps but in general less time is available on water-cooled reactors than on gas-cooled ones. Compare Section 3.3)

Other designs of reactor, still under development, are more 'user-friendly' than gas-cooled ones. The sodium-cooled fast reactor can remove heat from the core by natural circulation if external power supplies fail and if its temperature rises heat production falls off more rapidly than in other designs and ultimately becomes zero. In the High Temperature Gas

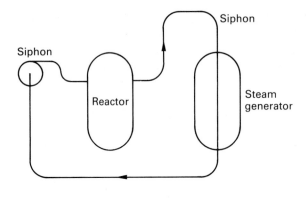

Figure 11.2(a) In the Three Mile Island design small differences in height and siphons prevent convective circulation

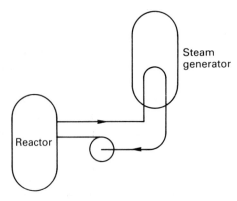

Figure 11.2(b) These difficulties are overcome in a modern design

reactor a small graphite-moderated core is cooled by high pressure helium. The high temperature resistance of the fuel and the high surface-to-volume ratio ensure that the after-heat is lost by radiation and conduction. In the Swedish Process Inherent Ultimate Safety (PIUS) Reactor a water-cooled core is immersed in a solution of boric acid in water. If the coolant pumps fail the boric acid solution is drawn through the core by convection. The boron absorbs neutrons and stops the chain reaction while the water removes the residual heat. No make-up water is needed for a week[20,21].

A halfway-house to inherently safer designs are those containing passive safety features. For example, in Three Mile Island and other early PWR designs the heat source, the core, is only a little below the heat sink, the steam generator, and so there is little or no convection cooling if the circulation pumps stop (Figure 11.2a). In addition, at Three Mile Island, siphon loops (high points in the piping) filled with steam and prevented flow and other obstructions increased the pressure drop. These problems are overcome in the modern design shown in Figure 11.2b. The steam generator is well above the core and there are no siphon loops. Convection cooling can supply 25% of the full cooling load[21].

The designers of Three Mile Island may have been familiar with advanced nucleonics but they did not have an instinctive grasp of the principles of flow, of what will go where without being pumped. Or perhaps they were being conservative and copying the first PWR designs which were intended for installation in submarines where all the equipment has to be on much the same level.

There are lessons here for the chemical industry. Chemical engineers do not always make as much use of convection cooling as they might. Safety apart, money can be saved if convection cooling can replace or assist pumped flow. Unfortunately it is easier to draw a pump on a flowsheet than juggle equipment to maximise convection. A new plant was a copy of an existing one except that the floors were 10 feet (3 m) apart instead of 8 feet (2.5 m). The increase was just enough to prevent convective flow[22].

An example of passive safety in the chemical industry is the use of insulation for fire protection instead of water cooling, which is active, that is, it has to be commissioned by people and/or equipment and they may fail to do so. Insulation does not have to be commissioned, it is ready for immediate use, but it is not inherently safe as bits may fall off or be removed for maintenance and not be replaced. The inherently safer solution is to use non-flammable materials, if we can, and remove the risk of fire.

A nuclear engineer opposed to inherently safer designs has written, 'building on what is already proven could bring swifter results with greater confidence than launching into radically new methods that purport to offer inherent safety'[23]. This sounds convincing until we remember that similar arguments could have been (and probably were) used 170 years ago to advocate the breeding of better horses instead of developing steam locomotives.

In the short term the Pressurised Water Reactor (PWR) with its complex replicated added-on safety systems may be the right answer for the UK but the advantages of the inherently safer designs are so great that we may be building PWRs for only a few decades. For those countries that lack the resources, culture or commitment necessary to maintain complex added-on safety systems PWRs are not the answer even today and they should wait until inherently safer designs are available.

11.8 Measure directly what we need to know

Not measuring what we need to know but instead measuring something from which it can be inferred (see Section 11.2) has caused accidents or operating problems in the process industries. Here are two examples.

In plants that handle flammable gases the control building is often pressurised to prevent the entry of flammable gas. Sparking electrical equipment can then be used in the building. An alarm sounds if the

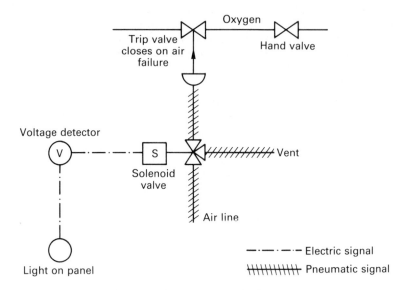

Figure 11.3 The light shows that the solenoid is de-energised, not that the oxygen flow has stopped

pressurising fails. The actual pressure in the building is low and difficult to measure so the voltage applied to the fan motor, or the current flowing through the motor, is sometimes measured instead. However, even though the motor is turning, the coupling to the fan may be broken, the impellor may be damaged or loose, there may be an obstruction in the ducting, such as a choked filter or a plastic sheet over the air inlet, or doors and windows may be open.

Ethylene oxide is manufactured by oxidising ethylene gas with oxygen. If a fault condition, such as a high oxygen level, occurs, a valve in the oxygen supply line is closed automatically. One day a light on the panel told the operator that this had happened. He knew the nature of the fault and that he would be able to start up the plant immediately so he did not close the hand valve on the oxygen line as well. Before he could restart the plant it exploded. The oxygen valve had not closed and oxygen had continued to enter the plant.

The valve was a pneumatic one and was closed by venting the air supply to the valve diaphragm, via a solenoid valve (Figure 11.3). The light on the panel merely said that the solenoid valve had been de-energised. Even though this occurs the solenoid valve may not move, the air may not be vented or the trip valve may not move. With some designs of valve the spindle can move without the valve closing. Actually, the air was not vented as the 1 inch vent line was plugged by a wasp's nest. The incident occurred in Texas where the wasps are probably bigger than anywhere else.

Event	Recommendations for prevention/mitigation

Development of US nuclear
industry halted

↑ ←———————————— **Provide information that will help public keep risks in perspective.**

Extreme public reaction

↑ ←———————————— **Plan how to handle public relations in emergencies.**

Damage to core

↑ ←———————————— Various design changes.

Water level fell

↑
Operators shut down water pumps

↑ ←———————————— **Pass on lessons of incidents elsewhere.**

Operators believed indicator light
and ignored other readings

↑ ←———————————— **Train operators in diagnosis. Provide diagnostic aids. Train operators to understand, not just to follow the rules. Consider minor failures as well as major ones.**

PORV stuck open but indicator
light said it was shut

↑ ←———————————— Measure directly what we need to know.

Primary water boiled
and PORV lifted

↑
Turbine tripped

↑
Instruments failed

↑
Water entered instrument air lines

↑ ←———————————— Inspect NRVs regularly, if possible (or use fluidic NRVs).

Attempt to clear choke
with instrument air

↑ ←———————————— Do not connect services to process lines at higher pressure. Do not use instrument air for line blowing.

Resin polisher unit choked

↑

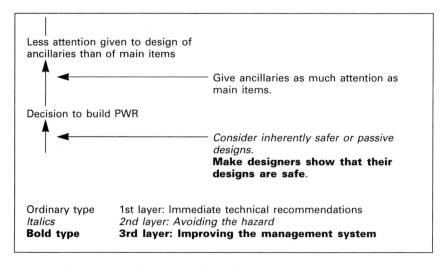

Figure 11.4 Summary of Chapter 11 – Three Mile Island

References

This chapter is based on an article which was published in *Hydrocarbon Processing*, Vol. 61, No. 6, June 1982, p. 187 and thanks are due to Gulf Publishing Co. for permission to quote from it.

1 Hudson, P. T. W., 'Psychological aspects of management for reduction of environmental risks', *Studiecentrum von Bedrief en Overheid*, Scheveningen, The Netherlands, 16–17 May 1988, p. 1.

2 Brooks, G. L. and Siddall., E, *An Analysis of the Three Mile Island Accident*, CNS First Annual Conference, Montreal, Canada, 18 June 1980.

3 Lewis, H. W., *Science*, March 1980, p. 33.

4 Lanouette, W. J., *The Bulletin of the Atomic Scientists*, January 1980, p. 20.

5 Schneider, A., *Loss Prevention*, Vol. 14, 1981, p. 96.

6 *Report of the President's Commission on the Accident at Three Mile Island (The Kemeny Report)*, Pergamon Press, New York, 1979.

7 *TMI-2 Lessons Learned Task Force Report*, Report No NUREG-0585, US Nuclear Regulatory Commission, 1979.

8 *Investigation into the March 28, 1979 Three Mile Island Accident by the Office of Inspection and Enforcement*, Report No 50-320/79-10, US Nuclear Regulatory Commission, 1979.

9 *Accident at the Three Mile Island Nuclear Power Plant: Oversight hearings before a task force of the Subcommittee on Energy and the Environment of the Committee on Interior and Insular Affairs of the House of Representatives*, Serial No. 96–8, 1979.

10 Moss, T. H. and Sills, D. L., (eds), *The Three Mile Island Nuclear Accident: Lessons and Implications*, Annals of the New York Academy of Science, Vol. 365, 1981.

11 *The Chemical Engineer*, No. 467, December 1989, p 28..

12 Reason, J., *Human Factors in Nuclear Power Operations*, Submission to a Sub-Committee of the House of Lords Select Committee on Science and Technology, undated.

13 Marshall, E. C. *et al.*, *The Chemical Engineer*, Feb. 1981, p. 66.

14 Duncan, K. D., *New Technology and Human Error*, edited by J. Rasmussen, K. D. Duncan and J. Leplat, Wiley, Chichester, UK, 1987, Chapter 19.

15 Andow, P. K., *Plant/Operations Progress*, Vol. 4, No. 2, April 1985, p. 116.

16 Kletz, T. A., *Lessons from Disaster – How Organisations have No Memory and Accidents Recur*, Institution of Chemical Engineers, Rugby, UK, 1993.

17 Kletz, T. A., *An Engineer's View of Human Error*, 2nd edition, Institution of Chemical Engineers, Rugby, UK, 1985, Section 4.2.

18 Hicks, D., *Process Safety & Environmental Protection*, Vol. 71, No. B2, May 1993, p. 75.

19 Franklin, N., *The Chemical Engineer*, No. 430, Nov. 1986, p. 17.

20 Weinberg, A. M. and Spiewak, I., *Science*, Vol. 224, No. 4656, 29 June 1984, p. 1398.

21 Forsberg, C. W. *et al.*, *Proposed and Existing Passive and Inherent Safety-Related Structures, Systems, and Components (Building Blocks) for Advanced Light-Water Reactors*, National Technical Information Service, Springfield, Virginia, especially pages 5–4 and 6–65.

22 Crowl, D. A., Private communication.

23 *Atom*, No. 394, August 1989, p. 35.

Chapter 12

Chernobyl

What are you worried about? A nuclear reactor is only a samovar.

The manager of a Russian nuclear power station[1].

It is, it seems, in the nature of human beings to seek scapegoats. In the aftermath of a major catastrophe there is an instinctive search for an individual, or individuals to blame.

A. Cruikshank[2]

Like the accident at Three Mile Island (Chapter 11) the accident at Chernobyl, in the Ukraine, then part of the USSR, on 26 April 1986 was the result of overheating of a water-cooled nuclear reactor and both accidents were said to be due to human error. Three Mile Island was clearly due to a failure to understand what was going on, the result of inadequate training. In the first edition of this book I said that Chernobyl was the result of a deliberate decision to ignore the normal safety instructions and override the normal protective equipment. While it is true that the normal safety equipment was overridden, the Russian authorities now admit that the safety instructions were not as clear as they should have been and that the operators may not have fully understood the consequences of their actions. There is little doubt that throughout the Russian design, operating and regulatory organisations the safety culture was deficient and that the first reaction, after the accident, was to look for culprits rather than look for the underlying weaknesses in the organisation, a common failing[2].

At Three Mile Island the discharge of radioactive material was too small to cause any harm. At Chernobyl about thirty people were killed immediately or died within a few months and it has been estimated that several thousand more will die from cancer during the next forty years[3]. (For comparison, about thirty million people will die from cancer in Europe during this period.) However, this figure is subject to considerable uncertainty. It assumes that the effects of small doses of radiation are proportional to the dose, an assumption that may not be correct, and that no

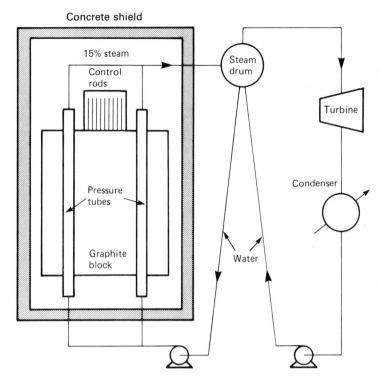

Figure 12.1 The Chernobyl reactor, simplified. The fuel is contained in 1661 zirconium alloy pressure tubes set in a graphite block. Water passes through the tubes. There are over 200 control rods

significant advances in the treatment of cancer will be made. We shall never know whether or not the estimate is correct as such a small increase (0.01%) will never be noticed.

Chernobyl released a million times more radioactivity than Three Mile Island but the atomic bomb tests of the late 1950s and early 1960s released a hundred times more than Chernobyl[12].

12.1   The reactor and the experiment

At Three Mile Island the water that cooled the reactor was kept under pressure and did not boil. In contrast the Chernobyl reactor was cooled by boiling water. About half the USSR reactors were of this type; the design is one that is not used outside the USSR, though other designs of boiling water reactor are used.

Figure 12.1 is a simplified diagram of the Chernobyl reactor. The uranium fuel is contained in 1661 pressure tubes, made of zirconium alloy, through which the water is passed. About 15% of the water is turned into steam which is separated from the water in a steam drum and drives a turbine. The exhaust steam is condensed and recycled. Note that there is only one water stream – not two, as in a pressurised water reactor. The pressure tubes are located in a large block of graphite which acts as a moderator, that is, it slows the neutrons down so that they will react with the uranium. (In a pressurized water reactor the water is the moderator; steam formation, should it occur, results in less moderation and less reaction. In the Chernobyl design the water layer is too thin to do much moderating but nevertheless absorbs neutrons; steam formation results in less absorption and thus more reaction.)

The graphite block is surrounded by a concrete shield, 2 m thick at the side and 3 m thick on top. About 200 control rods can move in and out of the reactor to absorb neutrons (and thus control the rate at which heat is produced) and to shut down the reactor in an emergency.

If reaction is stopped heat is still produced, initially at about 6% of the normal rate, by radioactive decay, so cooling has to be kept in operation or the reactor will overheat (as in a pressurised water reactor). As the reactor is not now producing power, the power to drive the water pumps and other auxiliary equipment has to come from the grid. Diesel generators are provided in case the grid supply is not available. They take about a minute to start up. During this time there may be no power available. The managers of the power station wanted to know if the turbine, as it slows down, will generate enough power to keep the auxiliaries on line during this minute. An experiment had shown that it would not but some changes had been made to the electrical equipment and they decided to repeat the experiment when the reactor was shutting down for a planned overhaul.

The experiment was a perfectly reasonable one to undertake and had been carried out before without incident but now we come to the fatal errors of judgement. The managers wanted to carry out the experiment several times. If they shut the reactor down it would take time to get it back on line, time that was not available, so they decided to keep the reactor on low rate and and simply isolate the steam supply to the turbine. The reactor was steadied out, with some difficulty, at 6% of full output although it should not have been operated at less than 20% output (for reasons which will be clear later) and it was intended to carry out the experiment at 25% output. An instruction forbidding operation at low outputs had not been issued[2], though it should have been, and there was no trip system to prevent operation at low outputs.

It took a long time to get the reactor steadied out at this low rate – the test had to be postponed as power was required urgently – and while operating under these abnormal conditions there was a build-up of xenon in the core. Xenon is formed during operation at low rates and is used up at high rates. It is a 'poison', as it absorbs neutrons and slows down the rate of nuclear fission, and therefore most of the control rods had to be withdrawn to keep the reactor running. In addition, because of the low rate, the cooling system contained few steam bubbles and the water absorbed more neutrons. As a result the equivalent of only six rods were fully inserted though at least thirty should always have been present. It was thought originally that this was due to a deliberate violation of a clear rule by the operators but it now appears that the computer which calculated the minimum number required was unreliable and that the operators did not understand why it was important, from a safety point of view, to keep a minimum number of rods inserted[2].

Finally the automatic trip system, which shuts down the reactor if there is a fault, was isolated. There were thus at least three departures from safe operation:

- Operating at 6% output instead of more than 20%.
- Inserting only the equivalent of six rods instead of thirty or more.
- Isolating the automatic trip system.

The emergency cooling system had to be isolated for the experiment to take place. It was isolated, with the approval of the chief engineer, 11 hours before the experiment started but not reinstated when the experiment was postponed. Although this did not affect the outcome of the accident in any way it does indicate the attitude towards safety on the plant[1].

12.2   The experiment goes wrong

The experiment was not successful. The turbine, as it slowed down, did not produce enough power to keep the water pumps running at full speed and the reactor temperature started to rise. The automatic trip should have dropped the control rods and stopped the nuclear reaction but it had been switched off. The operators tried to insert the control rods by manual control but they could not insert them quickly enough, the reactor temperature continued to rise, the heat production increasing a hundred times in one second, and the temperature reached 3000–4000°C.

The temperature rose so rapidly because at low output rates, below 20%, the reactor had a positive power coefficient, that is, as the temperature rose the rate of heat production also rose. (No other design of commercial reactor anywhere in the world has this feature.) For this reason the reactor should never have been operated below 20% load. In fact, it is doubtful if even the automatic equipment could have inserted the control rods quickly enough at low output rates. As the temperature rose, the water boiled more vigorously, more bubbles were formed and the heat transfer coefficient fell. The rising temperature in the fuel caused an increase in its internal pressure, as it contained volatile fission products, and this caused the fuel to disintegrate. The steam produced burst some of the pressure tubes, damaged the graphite and blew the 3 m thick concrete shield on the top of the reactor to one side. The building was damaged and fuel and fission products discharged into the atmosphere. About thirty people were killed immediately or died within a few weeks and, as already stated, many more will probably die in the years to come. Over 100 000 people were evacuated from an area 30 km in radius around the reactor and part of this area will be unfit to live in for many years. The fission products were detected in most parts of Europe.

12.3   The lessons to be learned

The most obvious lessons are *protective equipment should not be isolated and basic safety rules should be clearly stated and should not be ignored.* In addition, as the consequences of breaking the rules were so serious, *the plant should have been designed so that the rules could not be ignored*, that is, so that the automatic trip could not be isolated, so that at least thirty control rods had to be left in the core and so that the plant could not remain on line if the output was reduced below 20%. The Russian designers seem to have assumed that instructions would always be clearly stated and obeyed and that therefore there was no need for them to install protective equipment to prevent unsafe methods of operation. They assumed that those who issue and enforce operating rules and those who are supposed to follow them would be more reliable than automatic equipment, but they were not. (This is not a universal rule. Sometimes operators are more reliable than automatic equipment, but not in this case.)

Russians have a reputation for rule-based behaviour and for referring every detail to higher authority so it is at first sight surprising that the procedures at Chernobyl were so slipshod. The operating staff seem to have had contradictory instructions: to carry out the tests as quickly and effectively as possible and to follow normal operating procedures. They may have assumed that the instruction to carry out the experiment overrode the normal procedures. In the process industries managers have often had to argue with research workers who wanted to carry out experiments on the plant but did not want to be bound by the normal safety instructions. *Safety instructions should be followed at all times unless an exception has been authorised at the appropriate level after a systematic consideration of the hazards.* In addition, managers who talk a lot about output or efficiency or getting things done, without any mention of safety, inevitably leave operators with the impression that safety is less important. Managers should remember, when giving instructions, that *what you don't say is as important as what you do say.*

It seems that the local managers and perhaps the operators had been departing from the rules, or from good operating practice, for some time. Everything had to be referred to the top so it was necessary to break the rules in order to get anything done. Regular audits, if there had been any, would have picked up the fact that rules and good practice were not being followed but perhaps those at the top preferred not to know so that they would not be responsible. Holloway[13] cites an incident at another Russian nuclear power station when the director told the deputy premier responsible for energy production that the plant would not be ready on time as equipment was delivered late. The minister exploded, 'Who gave you the right, comrade, to set your own deadlines in place of the government's?' In this sort of climate people say nothing or tell lies. The Russians have no independent inspection service similar to the UK Nuclear Installations Inspectorate.

The early reports on Chernobyl did not make it clear at what level the decisions to operate at low rate, withdraw most of the control rods and switch off the trip were taken but we now know that the deputy chief engineer was present throughout the experiment[16]. It is clear, however, that when the normal procedures were suspended the operators had insufficient understanding of the process to be able to replace rule-based behaviour by skill-based behaviour. At Three Mile Island, when an unforeseen fault occurred, the operators were similarly unable to cope.

Rule-based behaviour is appropriate when straightforward tasks have to be performed but process plants do not come into this category. (In theory, if rule-based behaviour is all that is needed a computer can be used instead of a person.) On process plants we should never rely solely on rule-based behaviour as circumstances may arise, in fact probably will arise, which were not foreseen by those who wrote the rules. Note also that skill-based behaviour requires motivation as well as skill. The operators may not have understood why it was so important to operate above 20% output or why so many control rods should have been lowered. Like many operators, they may have relied more on process feel than theoretical knowledge and felt confident of their ability to control the reactor under all circumstances.

On process plants protective equipment sometimes has to be isolated, for example, if it goes out-of-order or if the plant is operating under abnormal conditions – a low flow trip may have to be isolated during start-up, for example – but they should be isolated only after authorisation at an appropriate level and the fact that the trip is isolated should be signalled in some way, for example, by a light on the panel (see Section 2.5). If possible the trip should reset itself after a period of time. When the potential consequences of a hazard are serious, as in a nuclear power station, isolation of a trip should be impossible. High reliability is then essential and it can be achieved by redundancy. Similarly, it is neither necessary nor possible to install protective equipment to prevent operators breaking every safety rule but certain modes of operation may have such serious consequences that it should be impossible to adopt them.

The managers do not seem to have asked themselves what would occur if the experiment was unsuccessful. *Before every experiment we should list all possible outcomes and their effects and decide how they will be handled.*

Underlying the above was a basic ignorance of the first principles of loss prevention among designers and managers and the comments made in

Chapter 10 on Bhopal apply even more strongly to Chernobyl. The Russians now agree that there were errors in the design of the reactor and they have made changes to their other reactors, for example, they have limited the extent to which control rods can be withdrawn and they are now kept in a region of high neutron flux so that their movement will have an immediate effect. In addition, they have made it harder for the protective systems to be defeated. They also agree that there were errors of judgement by the managers, who have been prosecuted and sent to prison (rather unfairly as the culture of the whole organisation was at fault), and they have agreed that the training of the operators was inadequate, but it is not clear that they recognise the need for changes in their approach to loss prevention or the need to answer questions such as those asked by Mosey[1]. He describes several institutional failings, as he calls them, but institutions have no minds of their own and I prefer to call them senior management failings.

- Why were the deficiencies in the Chernobyl design, well known in the Soviet nuclear industry as the result of incidents elsewhere[2], not addressed more thoroughly and conscientiously?
- Why was an electrical engineer, who was not a reactor specialist, put in charge of a nuclear power station?
- Why did the senior managers turn a blind eye to rule breaking?
- Why was the test programme not reviewed and approved?
- Why was there no effective regulatory or audit system?

One Russian who saw clearly what was wrong, was Valery Legasov, First Deputy Director of the Atomic Energy Institute, who killed himself two years after the accident[1]. He wrote[13]:

> ...the accident there was the apotheosis and culmination of all the improper management of the branch that had been going on for many decades... There was a total disregard of the point of view of the designer and the scientific manager, and it took a real fight to see that all the technological requirements were fulfilled. No attention was paid to the condition of instruments and the condition of equipment...

12.4   The need for inherently safer designs

The comments made in Chapter 11 on Three Mile Island apply with greater force to Chernobyl. In addition to the need for emergency cooling systems, which is common to all water-cooled reactors, the Chernobyl design was inherently less safe than the reactor at Three Mile Island. As already stated, if it got hotter the rate of heat production increased. In addition, at temperatures above 1000°C the water reacted with the zirconium pressure tubes, generating hydrogen. The hydrogen may have caught fire after the top was blown off the reactor. Once the pressure tubes burst the water may have reacted with the graphite, also producing hydrogen, but this is an endothermic reaction and therefore less serious than the zirconium–water reaction. Whenever possible we should *avoid using in close proximity materials which react with each other* even though reaction does not occur until the temperature rises, a container leaks or there is some other departure from normal operating conditions (compare Section 10.3 on Bhopal).

Another weakness of the Russian reactor was its susceptibility to knock-on effects. Once a number of pressure tubes burst – considered an 'incredible' event by the designers – other tubes and the graphite were damaged and the concrete top was blown off the reactor. The roof of the building was flammable and caught fire.

12.5   Public response

Chernobyl was by far the worst nuclear accident that has occurred, though comparable with Bhopal if the probable long-term results are taken into account, and the public response was correspondingly great. The official attitude in the West was that the design of the reactor, and licensing procedures, were so different from those in use elsewhere that there was no reason to doubt the safety of reactors in other countries but the public and press remained sceptical. Though the public is inclined to treat the nuclear industry as one, it does not treat other industries in the same way. Collapse of a dam in India does not make people doubt the safety of UK dams or result in demands for them to be phased out. Perhaps the difference is due to the fact that the discharge from Chernobyl spread across national boundaries. (London is more exposed to fall-out from French and Belgian reactors than from UK ones.) While the detailed lessons of Chernobyl have been studied in the West there does not seem to have been any major change in designs or procedures, legislative or otherwise. In this respect it has had less effect than Flixborough, Seveso or Bhopal. However, in the long run Chernobyl may encourage the development of inherently safer or more user-friendly reactors, as discussed in Chapter 11, and the need for the nuclear industry to sell itself to the public is now greater than ever.

For some time after the accident people in the West were asking the Russian government to shut down all reactors of the Chernobyl design but we now realise that this would have serious economic and social effects and instead Western governments are helping the Russians, and other former Communist countries, to improve their designs and methods of operation[14], for all designs of reactor[15].

We should keep the widespread effects of Chernobyl in perspective. If a UK smoker who is worried about the effects smokes one cigarette less – not one per day or per year but one once and for all – he will have reduced his risk of getting cancer more than Chernobyl has increased it.

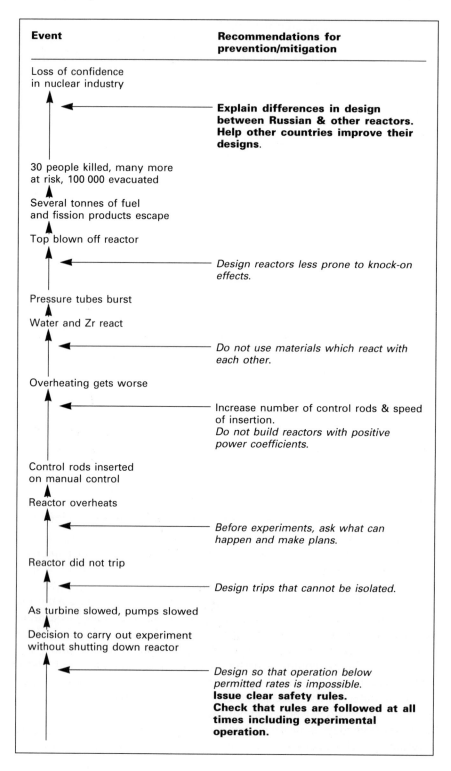

Event	Recommendations for prevention/mitigation

Loss of confidence in nuclear industry

← **Explain differences in design between Russian & other reactors. Help other countries improve their designs.**

30 people killed, many more at risk, 100 000 evacuated

Several tonnes of fuel and fission products escape

Top blown off reactor

← *Design reactors less prone to knock-on effects.*

Pressure tubes burst

Water and Zr react

← *Do not use materials which react with each other.*

Overheating gets worse

← Increase number of control rods & speed of insertion.
Do not build reactors with positive power coefficients.

Control rods inserted on manual control

Reactor overheats

← *Before experiments, ask what can happen and make plans.*

Reactor did not trip

← *Design trips that cannot be isolated.*

As turbine slowed, pumps slowed

Decision to carry out experiment without shutting down reactor

← *Design so that operation below permitted rates is impossible.*
Issue clear safety rules.
Check that rules are followed at all times including experimental operation.

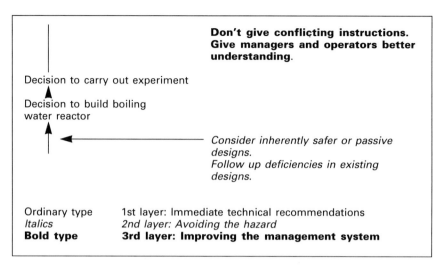

Decision to carry out experiment

Decision to build boiling
water reactor

**Don't give conflicting instructions.
Give managers and operators better
understanding.**

*Consider inherently safer or passive
designs.*
*Follow up deficiencies in existing
designs.*

Ordinary type	1st layer: Immediate technical recommendations
Italics	*2nd layer: Avoiding the hazard*
Bold type	**3rd layer: Improving the management system**

Figure 12.2 Summary of Chapter 12 – Chernobyl

Arrangements in the UK for monitoring fall-out and its accumulation in food and for informing and advising the public were found to be lacking and many changes have been made in this area.

12.6 Appendix

While this book was in production a new and very readable account of Chernobyl appeared (*Ablaze – The Story of Chernobyl*, by P.P. Read)[16]. It is not strong on the technical causes of the disaster and adds little to what we already know but it does bring out very clearly the following underlying weaknesses in the safety culture and management system:

(1) Supply problems led to improvisation on site and poor working and living conditions that did not attract skilled and experienced workers (pp 40, 61 and 68).
(2) The director of the power station had many outside responsibilities (pp 49, 74).
(3) Patronage was rampant and the chief engineer was a political appointee in poor health (pp 53, 58, 60 and 73).
(4) Accidents elsewhere were hushed up. Employees did not even know what had happened on the other reactors on the same site (pp 53–57).
(4) There was a belief that if the regulations were followed nothing could go wrong (p 54).
(5) Information display was poor (p 54).
(6) Rules had to be broken to start the reactor on time and to keep it running (pp 53 and 64).

Read's book also shows that at the time of Chernobyl Gorbachev's new policy of *glasnost* (openness) was still in its early days and the Soviet government was not willing to admit that there was anything wrong with their designs or safety culture. The accident had therefore to be blamed on human error and the last of the Stalinist show trials was organised. In time errors were admitted (see the quotation from Legasov on page 126) and those sent to prison were released (pp 264–273).

References

The text is based mainly on references 1–11:
1 Mosey, D., *Reactor Accidents*, Butterworth-Heinemann, Oxford, 1990, p. 81.
2 Cruikshank, A., *Atom*, No. 427, March/April 1993, p. 44.
3 *Atom*, No. 366, April 1987, p. 16.
4 Gittus, J. H. and Dalglish, J. *Atom*, No. 360, Oct. 1986, p. 6.
5 Gittus, J. H. *Atom*, No. 386, June 1987, p. 2.
6 Franklin, N. *The Chemical Engineer*, No. 430, Nov. 1986, p. 17.
7 *Physics Bulletin*, Vol. 37, No. 11, Nov. 1986, p. 447.
8 Hohenemser, C., Deicher, M., Ernst, A., *et al.*, *Chemtech*, Oct. 1986, p. 596.
9 Rowland, P. R., *Chemistry in Britain*, Vol. 22, No. 10, Oct. 1986, p. 907.
10 Gittus, J. H., Hicks, D., Bonell, P. *et al.*, *The Chernobyl Accident and its Consequences*, Her Majesty's Stationery Office, London, 1987.
11 Reason, J., *Bulletin of the British Psychological Society*, Vol. 40, 1987, p. 201.
12 Grimston, M., *Atom*, No. 413, May 1991, p. 12.
13 Holloway, D., *Chemtech*, Feb. 1991, p. 82.
14 Tyror, G., *Atom*, No. 423, July/August 1992, p. 10.
15 Laaksonen, J., *EQE Review*, Fall 1992, p. 1.
16 Read, P.P., *Ablaze – The Story of Chernobyl*, Secker and Warburg, London, 1993.

Chapter 13

Aberfan

They take back bright to the coiner the mintage of man.

A. E. Housman, *A Shropshire Lad*

On 21 October 1966 a waste tip from a coal mine collapsed onto the village of Aberfan in South Wales. A school lay in the path of the waste and the 144 people killed were mainly children. Compared with the other accidents described in this book the immediate technical causes of the disaster were simple – the tip was constructed above a stream on sloping ground – but the official report[1] brought out very clearly the underlying weaknesses in the management system, namely, a failure to learn from the past, a failure to inspect regularly and a failure to employ competent and well-trained people with the right professional qualifications. The official report, on which this chapter is based, also includes many perceptive remarks which apply to accidents generally. Quotations from the report are in *italics* and the references are to paragraph numbers.

13.1 A failure to learn from the past

. . .forty years before it occurred, we have the basic cause of the Aberfan disaster being recognised and warned against. But, as we shall see, it was a warning which went largely unheeded (§44).

In 1927 Professor George Knox presented a paper on 'Landslips in South Wales Valleys' which became widely known at the time. He gave full warning of the menace to tip stability presented by water and said that 'if you do not pay for drainage, you will have to pay for landslips in other ways'.

Tip slides are not new phenomena. Although not frequent, they have happened throughout the world and particularly in South Wales for many years, and they have given rise to quite an extensive body of literature available long before the disaster (§72). In 1939 there was a tip slide at

Cilfynydd in South Wales. *Its speed and destructive effect were comparable with the disaster at Aberfan, but fortunately no school, house or person
lay in its path. . . . It could not fail to have alerted the minds of all reasonably prudent personnel employed in the industry of the dangers lurking in
coal-tips . . . the lesson if ever learnt, was soon forgotten* (§82).

*In 1944 another tip at Aberfan slid 500–600 feet. Apparently, no-one
troubled to investigate why it had slipped, but a completely adequate explanation was to hand ... it covered 400 feet of a totally unculverted stream*
(§88). *To all who had eyes to see, it provided a constant and vivid reminder
(if any were needed) that tips built on slopes can and do slip and, once
having started, can and do travel long distances* (§89).

*Why was there this general neglect? Human nature being what it is, we
think the answer to this question lies in the fact that . . . there is no previous case of loss of life due to tip stability . . .* (§68). The Inspectorate of
Mines and Quarries were not even informed of the 1944 slide as no-one
was killed or injured and they never looked on tip stability as a problem
meriting close inspection or recommendation.

Aberfan therefore shows very clearly that we should learn from all
accidents, those that could have caused death or injury as well as those that
did, and that a conscious management effort is needed to make sure that the
lessons of the past are not forgotten. Reference 2 describes many other
accidents that were forgotten and then repeated, often in the same company.

13.2   A failure to inspect adequately

*The simple truth is that there was no regular inspection of the tips. On the
contrary, their inspection (such as it was) was wholly haphazard in point
of time and had no reference to their stability at all, but simply related to
such matters as the condition of the mechanical equipment for tipping* (§64).

In 1965 there was a tip slide at Tymawr Colliery in South Wales. A main
road was flooded and two or three cars in the colliery car park were
damaged. The Chief Engineer of the South Western Division of the
National Coal Board sent a memorandum, 'Precautions to Prevent
Sliding', an update of one issued after the 1939 slide, to the Area Chief
Engineers, Mechanical Engineers and Civil Engineers under his control.
In the covering letter he said, '*I should be pleased, therefore, if you would
arrange with your colleagues, for a detailed examination of every tip within
your Area, and to take the necessary action for its immediate safety and
ultimate good management*' (§162).

*The inspection of the Aberfan tips was carried out by the Area Mechanical
Engineer. . . . it was remarkably inadequate. It was purely visual; (he) had
no plans or surveys with him and he made no notes . . . and he never climbed
to the top of Tip 7 to see conditions for himself or to seek information from
the tipping gang . . . There was in truth much for him to note – the extent of
the 1944 slip, the results of the incident of 1963, and the deep bowl-like*

depression, to mention just some of the unusual features ... Had these obvious features been noted and properly enquired into, it seems inconceivable that there would have been a disaster (§170).

13.3   A failure to employ competent and well-trained people

The Area Mechanical Engineer was strongly criticised by the Enquiry *(for failing to exercise anything like proper care in the manner in which he purported to discharge the duty of inspection laid upon him* [§171]) but he was a mechanical engineer, not a civil engineer, and had had no training to qualify him for the task of tip inspection. In addition, like the engineers who built the pipe which failed at Flixborough (Chapter 8), he had no professional training or qualification but had started as an apprentice at the age of 15. Like them he did not know what he did not know.

It was customary in the coal industry for tips to be the responsibility of mechanical rather than civil engineers. *It was left to the mechanical engineer to do with tips what seemed best to him in such time as was available after attending to his multifarious other duties (§70). For our part, we are firmly of the opinion that, had a competent civil engineer examined Tip 7 in May, 1965, the inevitable result would have been that all tipping there would have been stopped (§73). The tragedy is that no civil engineer ever examined it and that vast quantities of refuse continued to be tipped there for another 18 months* (§168).

Why did the Area Civil Engineer, who also received the 1965 memorandum, not take an interest in the tip? It was *traditional that tips were the responsibility of mechanical engineers.* In addition, the Civil Engineer was heavily overshadowed by the far more dominant Mechanical Engineer and took it for granted that his assistance was not wanted (§165).

Tip 7 was started at Easter 1958 ... The two men who took the decision to start it were ... *quite unfitted by training to come to an unaided decision as to the suitability of the proposed site ... They had no Ordnance Survey map, and they took no plan with them, because none existed; they made no boreholes; they came to no conclusion regarding the limits of the tipping area; and they consulted no one else – not even the Colliery Surveyor. They arranged for no drainage, for they considered none necessary. It was a case of the blind leading the blind ... in a system which had been inherited from the blind (§104). In our judgment, such inspection as they made was worthless. They were unfitted by training to judge the matter, and what stared them in the face they ignored* (§105).

13.4   Personal responsibility

We found that many witnesses, not excluding those who were intelligent and anxious to assist us, had been oblivious of what lay before their eyes. It did not enter their consciousness. They were like moles being asked about the habits of birds (§17).

... the Report which follows tells not of wickedness but of ignorance, ineptitude and a failure in communications. Ignorance on the part of those charged at all levels with the siting, control and daily management of tips; bungling ineptitude on the part of those who had the duty of supervising and directing them; and failure on the part of those having knowledge of the factors which affect tip stability to communicate that knowledge and to see that it was applied (§18).

The stark truth is that the tragedy of Aberfan flowed from the fact that, notwithstanding the lessons of the recent past, not for one fleeting moment did many otherwise conscientious and able men turn their attention to the problem of tip stability ... These men were not thinking or working in a vacuum. All that was required of them was a sober and intelligent consideration of the established facts (§46).

... the Aberfan disaster is a terrifying tale of bungling ineptitude by many men charged with tasks for which they were totally unfitted, of failure to heed clear warnings, and of a total lack of direction from above. Not villains, but decent men, led astray by foolishness or by ignorance or by both in combination, are responsible for what happened at Aberfan (§47).

The incidents preceding the disaster ... should, in our judgement, have served ... to bring home vividly to all having any interest in coal-mining that tips placed on a hillside can and do slip and, having started, can move quickly and far; that it was accordingly necessary to formulate and maintain a system aimed at preventing such a happening; and for that purpose to issue directions, disseminate information, train personnel, inspect frequently and report regularly. These events were so spread out over the years that there was ample time for their significance to be reflected upon and realised and so to lead to effective action. But the bitter truth is that they were allowed to pass unheeded into the limbo of forgotten things (§48).

... most of the men whose acts and omissions we have had to consider have had, as it were, a bad upbringing. They have not been taught to be cautious, they were not made aware of any need for caution, they were left uninformed as to the tell-tale signs on a tip which should have alerted them. Accordingly, if in the last analysis, any of them must be blamed individually for contributing to the disaster ... a strong 'plea in mitigation' may be advanced ... personal responsibility is subordinate to confessed failure to have a policy governing tip stability ... It is in the realm of an absence of policy that the gravest strictures lie, and it is that absence which must be the root cause of the disaster (§182).

Change a few words and the passages just quoted would apply to many more of the accidents discussed in this book, and many others.

The local Member of Parliament admitted at the Inquiry that he entertained the thought that the tip 'might not only slide but in sliding reach

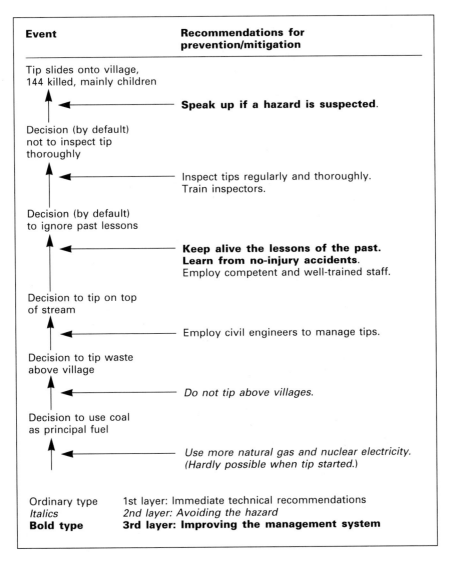

Figure 13.1 Summary of Chapter 13 – Aberfan

the village' but did nothing as he feared that the result of drawing attention to the tip might be closure of the colliery (§61). He presumably did not realise how serious the consequences of a slide might be but nevertheless it reminds us that anyone, especially someone in a position of authority, should speak up if he sees a hazard. If we see a hole in the road, then morally (though not legally, . . . *the law recognises no general duty to protect another from harm* [§77]) we are obliged to prevent other people falling in it (see Section 8.4) .

13.5   Inherently safer designs

Chapters 10–12 have stressed that whenever possible we should try to avoid hazards rather than control them. This is not discussed in the official Report but applying that philosophy leads to the suggestion that whenever possible we should locate tips so that, if they do slide, casualties are likely to be small.

Going further, what you don't have, can't slide. The less coal we use, the fewer tips we shall need. Increased use of natural gas and nuclear electricity will decrease our need for coal. We can hardly blame the National Coal Board for not pointing this out, especially as most of the tips in existence were started before natural gas and nuclear electricity became available. However, the opponents of nuclear electricity make much of the hazards of nuclear waste but the quantities involved are tiny compared with the waste produced by coal-mining. (If all electricity was made from nuclear energy, instead of 20% as in the UK, then each person in his lifetime would account for a piece of highly-active waste the size of an orange.) They do not remind us of Aberfan but it is just as relevant to point out what actually happened at Aberfan as what might happen in the disposal of nuclear waste. Of course, coal-mining tips can be managed safely, but so can nuclear waste disposal.

References

1 *Report of the Tribunal appointed to inquire into the Disaster at Aberfan on October 21st, 1966*, Her Majesty's Stationery Office, London, 1967. Thanks are due to the Controller of Her Majesty's Stationery Office for permission to quote from this report.
2 Kletz, T. A., *Lessons from Disaster – How Organisations have No Memory and Accidents Recur*, Institution of Chemical Engineers, Rugby, UK, 1993.

Chapter 14

Missing recommendations

Up to this moment I have stated, so far as I know, nothing but well-authenticated facts, and the immediate conclusions which they force upon the mind. But the mind is so constituted that it does not willingly rest in facts and immediate causes, but seeks always after a knowledge of the remoter links in the chain of causation.

Thomas H. Huxley, *On a Piece of Chalk*, 1868

We should include in accident reports all the facts that have come to light, even though no conclusions are drawn from some of them, so that readers with different backgrounds, experience or interests can draw additional conclusions which were not obvious to the original investigators.

For example, the official report on Flixborough (Chapter 8), though outstanding in many ways, did not discuss inherently safer designs. But the report did contain the information from which others could draw an important additional conclusion: the most effective way of preventing similar accidents in the future is to design plants which do not contain such large inventories of hazardous materials.

This chapter summarises three accident reports from which some readers were able to draw conclusions not apparent to the original authors. Two of the reports are United Kingdom official ones. This does not imply that the authors of official reports are poor at drawing conclusions, they are not, but they do take care to include information from which readers may be able draw additional conclusions.

Perhaps the most extreme example of a report which failed to draw the necessary conclusions was one which said that there was no need to recommend any changes in design as the plant had been destroyed!

14.1 A tank explosion

A hydrocarbon storage tank had to be gas-freed for entry. Air was blown into it with a fan. The vapour/air mixture in the tank had to pass through the explosive range. The fan impellor disintegrated and the resulting sparks ignited the vapour/air mixture. An explosion occurred.

The recommendation was to use a better fan.[1]

Most readers will have drawn the conclusion that tanks should be gas-freed with nitrogen, steam or other inert gas. If air is used an explosive mixture will be formed and experience shows that sooner or later a source of ignition will turn up. Using a better fan merely ensures that one particular source of ignition is less likely to turn up again. It does not deal with the underlying problem, that flammable mixtures are potentially hazardous and should never be permitted except under rigidly defined conditions where the risk of an explosion can be accepted.

14.2   The unforeseen results of technical change

Eleven men were killed in an explosion at a steelworks in Scunthorpe, Lincolnshire in 1975. Molten iron at 1500°C was being run out of a blast furnace into a closed ladle or torpedo, a long, thin vessel mounted on a railway truck and able to carry 175 tonnes of iron. The molten iron entered the torpedo through a hole in the top, 0.6 m in diameter. About 2–3 tonnes of water from a leak also entered the torpedo and rested on top of the molten iron, separated from it by a thin crust of iron and slag. When the torpedo was moved, the water came into direct contact with the molten iron and vaporised with explosive speed and violence. Ninety tonnes of molten iron were blown out and the pouring spot, weighing over a tonne, was blown onto the roof of the building.

A flow of water onto the top of molten iron in an old-fashioned open ladle did not matter as when it turned to steam there was plenty of room for the steam to escape. No-one realised, when the design of the ladle was changed, that entry of water would now be very dangerous.

The official report[2] described the cause of the water leak in detail – for twenty years plugs in the cooling system had been made of steel instead of brass and the report includes many photographs of corroded plugs – but this is irrelevant. If anyone had realised that water was dangerous the flow of water from the leak could have been dammed or diverted.

Both the official report and the company report, which it quotes, made a series of recommendations designed to prevent water leaks. They also recommended that torpedoes should not be moved until any water in them has evaporated and that better protective clothing should be provided. The underlying cause of the explosion, however, was the failure to see the result of technical change. The weakness in the management system was the lack of any formal or systematic procedure for examining technical changes in order to see if there were any unforeseen side-effects. The same weakness was present at Flixborough (Chapter 8) and in the accident discussed in Chapter 7.

The official report does quote the Safety Policy statement of the organisation which required 'a progressive identification of all hazards involving injury and/or damage potential'. The official report concluded that 'senior management had not implemented the declared safety policy' and that they 'should take urgent and comprehensive action to implement it'. However, this advice is not very helpful. How do we identify all hazards? No advice was given.

For major modifications hazard and operability studies[3], referred to several times already in this book, are now widely recommended. Simpler techniques have been described for minor modifications[4]. Both hazard and operability studies and the simpler techniques have been developed with the process industries in mind and may require some modification before they can be applied to industries such as steel production.

14.3   An explosion in a pumping station

In 1984 an explosion in the valvehouse of a water pumping station at Abbeystead, Lancashire killed sixteen people, most of them local people who were visiting the station. Water was pumped from one river to another through a tunnel. When pumping was stopped some water was allowed to drain out of the tunnel and leave a void. Methane from the rocks below accumulated in the void and, when pumping was restarted, was pushed through vent valves into an underground valvehouse where it exploded.

If the presence of methane had been suspected, or even considered possible, it would have been easy to prevent the explosion by keeping the tunnel full of water or by discharging the gas from the vent valves into the open air. In addition, smoking, the probable source of ignition, could have been prohibited in the valvehouse (though we should not rely on this alone; as mentioned in Section 14.1 and in Chapter 4 we should always try to prevent formation of a flammable mixture). None of these things were done because no-one realised that methane might be present. Although there were references to dissolved methane in water supply systems in published papers, they were not known to engineers concerned with water supply schemes.

The official report[5] recommends that the hazards of methane in water supplies should be more widely known but this will prevent the last accident rather than the next. Many more accidents have occurred because information on the hazards, though well known to some people, was not known to those concerned. The knowledge was in the wrong place. Another example was described in Chapter 4. Hazard and operability studies (hazops) will prevent these accidents only if a member of the hazop team knows of the hazard. Perhaps engineers should be encouraged to read widely and acquire a rag-bag of bits of knowledge that might come in useful at some time. At Abbeystead, if a hazop had been carried out, and if one member of the team had known, however vaguely, that methane might be present, and had expressed his misgivings, then the accident could have been prevented. It was not necessary for him to know the precise degree of risk or exactly what action to take; it would be sufficient for him to suspect that methane might be present and alert his colleagues.

Two significant points are not considered in the official report. The first is that an underground building is a confined space and many people have been overcome in such spaces. In particular carbon dioxide has often accumulated in underground confined spaces, with fatal results. Confined spaces in which there is any reason to suspect the presences of hazardous gas or vapour should be entered only after testing of the atmosphere, formal consideration of the hazards, gas-freeing and isolation (if necessary) and authorisation by a competent person. In the UK this is required by the Factories Act, Section 30.

The second point is the reason why the vent discharged into the valve-house, an unusual place for a vent discharge. The report does state (in paragraph 2) that as the area was one of 'outstanding scenic beauty ... everything possible should be done to minimise the effect of the scheme on the landscape' and for this reason the valvehouse was constructed underground. It seems that the vent pipe was routed into the valvehouse for the same reason. The explosion is thus another example of a design made with the admirable intention of protecting the environment which had unforeseen effects on safety[6] (see also page 33).

Three years earlier another explosion had been caused by a vent discharging into a building. A small factory recovered solvent by distillation. The cooling water supply to the condenser failed and hot vapour was discharged into a building where it exploded, killing one man, injuring another, seriously damaging the factory and causing some damage to surrounding houses. Again, the reason for the extraordinary location of the vent pipe discharge is not stated. It may have been routed into the building to try to minimise smells which had produced some complaints[7].

References

1   Kletz, T. A., *Loss Prevention*, Vol. 9, 1975, p. 65.
2   Health and Safety Executive, *The Explosion at the Appleby-Frodingham Steelworks, Scunthorpe on 4 November 1975*, Her Majesty's Stationery Office, London, 1976.
3   Kletz, T. A., *Hazop and Hazan – Identifying and Assessing Process Industry Hazards*, 3rd edition, Institution of Chemical Engineers, Rugby, UK, 1992.
4   Kletz, T. A., *Chemical Engineering Progress*, Vol. 72, No. 11, Nov. 1976, p. 48.
5   Health and Safety Executive, *The Abbeystead Explosion*, Her Majesty's Stationery Office, London, 1985.
6   Kletz, T. A., *Process Safety Progress*, Vol. 12, No. 3, July 1993, p. 147.
7   Health and Safety Executive, *The Explosion and Fire at Chemstar Ltd on 6 September 1981*, Her Majesty's Stationery Office, London, 1982.

Chapter 15

Three weeks in a works

I have walked round many factories in my 35 years with ICI and although I have not had in-depth experience of working on the shop floor, I can tell you that I can recognise pretty quickly the difference between a really good factory and one that has been tidied up specifically for my visit.

Sir Denys Henderson, Chairman of ICI

The accidents described so far in this book have been serious or unusual, usually both. By concentrating on these have we been giving a wrong impression and making recommendations that will have little impact on the majority of accidents?

To try to answer this question I have, on several occasions, with the help of colleagues, investigated every accident that occurred in a works during a three-week period. This chapter describes the results of one such investigation. I describe the accidents first and then make some general observations. We shall see that, as with the major incidents described so far, we can learn much more if we look beyond the immediate technical recommendations for ways of avoiding the hazards or for ways of improving the management system.

Ferry[1] has also examined some simple accidents and come to the same conclusion.

During the three weeks we devoted more time to the investigation of the accidents than might be practicable in the ordinary way. Nevertheless the results showed that the investigations normally carried out were often superficial and that there was more to be learnt from the accidents than was in fact being learnt. In addition, conclusions could be drawn from the accidents as a whole that could not be drawn from individual incidents. Information bought at a high price was not being used.

The works employed about one thousand people, including maintenance workers; apart from a few clerical and laboratory workers, they were all male. Half the works consisted of large continuous plants making a few chemicals in large tonnages while the other half consisted of batch plants making a range of small tonnage products, many of which were corrosive.

An accident was defined as any incident that resulted in attendance at the ambulance room for first-aid treatment. Many of the accidents were therefore very trivial, for example, dust in eye. Only one accident resulted in lost-time. We also investigated dangerous occurrences, that is, incidents that caused damage to plant or loss of product or raw material. Altogether there were thirty-two injuries and five dangerous occurrences (one of which caused injury) during the three weeks. We visited and photographed the scene of every accident and spoke to the injured men and to other people who were involved and we obtained copies of the accident reports. A hand-written report was started by the foreman and completed by the manager for every incident. In addition, for the more serious incidents, including all chemical burns and all dangerous occurrences, a more extensive typed report was produced and circulated to the senior management and the company safety adviser. These typed reports were models of their kind; the descriptions were crisp but no relevant details were omitted and clear recommendations for action were made and accepted. In contrast, the hand-written reports were often superficial and some of them did not arrive at the safety foreman's office until weeks later. A few never arrived at all, despite prompting by the safety foreman who had the unenviable job of trying to extract them from foremen and managers who did not see the need to fill in a form every time someone got dust in his eye. However, as we shall see there is something to be learnt from almost every incident.

Nine of the accidents were due to dirt or liquid entering the eye. Most of them occurred in circumstances where men would not normally be expected to wear goggles. The only effective method of prevention would be for every employee to wear safety spectacles at all times, but the management of the works did not feel that the effort required to persuade everyone to do so could be justified. An attempt had been made, with some success, to persuade workshop employees to wear safety spectacles and for some years all laboratory workers had been required to wear them. As they were staff employees it was easier to make rules and enforce them.

One company claims to have reduced eye injuries from four per week to about three per year by persuading all 150 employees, and contractors, to wear safety spectacle at all times. They were given a number of types to choose from[2].

Although there are a few criticisms of the works in the following pages, nevertheless its accident record was good and on the whole the attitude to safety was above average. At the time of the survey the works was well on the way to completing a million hours without a lost-time accident.

15.1   The accidents

The incidents are described in the order in which they occurred.

Accident 1

A process operator on one of the batch plants, brushed his forehead with his arm and some dirt from his overalls entered his eyes.

According to the report he was told to take more care, a piece of accident prevention liturgy that we will meet again and again. The report also recommended that after carrying out dirty jobs, men should change into clean overalls. It did not discuss the practicality of this proposal. How often would they have to change? Are the overalls available? It might have been better to have looked round the plant to see if overalls generally were dirty. In general, when investigating accidents we need to know whether we have uncovered an isolated incident or if there is an ongoing and unsatisfactory state of affairs.

Accident 2

Half-an-hour later the *same man* dipped a small tank manually, using a dip-stick. His goggles steamed up so he lifted them up to read the dip and then forget to replace them. As he was replacing the cover on the dip-hatch a splash of liquid entered his eye. He was again told to take more care.

Eye injuries were discussed above. The liquid in the tank was very viscous but even so it ought to be possible to provide a level measuring instrument and thus avoid the situation that lead to the accident.

Accident 3

A laboratory assistant complained of sore eyes. He may have rubbed them with contaminated hands, though he normally wears gloves. As the cause was uncertain no recommendations could be made.

Accident 4

The first dangerous occurrence: liquid was pumped intermittently from an atmospheric pressure storage tank to one which was kept at a gauge pressure of 1 bar. The pump suction and delivery valves were kept open and a non-return valve was relied on to prevent backflow (Figure 15.1).

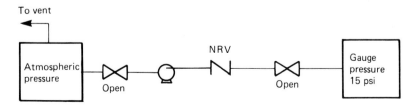

Figure 15.1 When the NRV failed, a leak was inevitable (see item 4)

A piece of wire, about 3 inches long, became stuck in the non-return valve and the atmospheric pressure storage tank overflowed. Two tonnes of liquid were spilt.

The operators claimed that they had used this method of operation for seven years without incident. However, non-return valves are always liable to fail and a spillage in the end was almost inevitable. The incident is a classic example of an accident waiting to happen.

Note that seven years experience without a spillage does not prove that the method used is safe. All it tells us is that we can be 86% confident that the chance of a spillage is once in 3½ years, or less[3].

Accident 5

While a welder was burning a hole in a pipe, 6 inches (150 mm) diameter, with walls 1/2 inch (12 mm) thick, in a workshop, a sudden noise made him jerk, his gun touched the pool of molten metal and a splash of metal hit him on the forehead. If the wall thickness had been above 1/2 inch a hole would have been drilled in the pipe first and the foreman said, on the accident report, that this technique should be used in future for all pipes. However, this would be troublesome. A crane would be needed to move the pipe to the drilling bay and back and this could cause delay. In practice, nothing was done.

The welder had cut thousands of holes before without incident, the chance of injury was small and it may have been reasonable to do nothing and accept the slight risk of another incident. In practice this is what the foreman, and the engineer in charge, decided to do. However, they were not willing to say so and instead they recommended action that they had no intention of enforcing.

After an accident many people feel that they have to recommend something even though they know it will be difficult and they have no intention of carrying out the recommendation with any thoroughness. It would be better to be honest and say that the chance of another incident is so low that the risk should be accepted. The law in the UK does not expect us to do everything possible to prevent accidents, only what is 'reasonably practicable'. Whether or not that was the right decision in the present case is a matter of opinion.

Accident 6

A small leak of liquefied petroleum gas (LPG) occurred from a drain valve on a pump suction line. It was detected by flammable gas detectors which were permanently installed nearby and was stopped by closing the valve before the gas ignited.

The drain valve was far below the company standards: it was a brass valve, of a type stocked for use in domestic water systems, it was connected by screwed joints, was inadequately supported, and was a single valve though the company standard called for two valves, 1 m apart (or one valve and a blank if the drain point is used only occasionally). The works had spent a substantial sum of money upgrading the standard of its LPG installations, the programme had been completed less than a year before and it is not known how the sub-standard valve came to be introduced. The incident showed up weaknesses in the system for controlling plant modifications. It was far too easy for any foreman who wanted to do so, to alter the plant.

A foreman entered the cloud of vapour (about 2 m across) to close the valve. Fortunately it did not ignite or he would have been seriously hurt. He should not have done so unless protected by water sprays. Perhaps practice in closing valves in this way should be included in fire training.

This incident resulted in three separate actions at company level:

(a) Improvement in the methods for controlling plant modifications. Other incidents, including the Flixborough explosion (Chapter 8) which occurred soon afterwards, emphasised the need for such improvements.

(b) An effort to discourage people from entering clouds of leaking gas to isolate leaks. I would not go so far as to say that no-one should ever do so, there may be occasions when, by taking a change for a few moments, someone has prevented a leak developing and ultimately exploding, but such cases should be exceptional. We should try to avoid putting people in situations where they have to make such decisions. Remotely operated emergency isolation valves should be installed on the lines leading to equipment which experience shows is liable to leak (see Section 4.2).

(c) A survey of other plants to see if there were any more sub-standard drain valves. Several were found.

If I had not been involved in the 'Three Weeks in a Works' exercise I might not have realised the significance of this incident and recommended the three actions above.

Accident 7

During fog a motor-cyclist hit a curb and came off. Though he was on company land the accident did not count towards the works record as he was on his way home and for this reason very little interest was taken in it, though it was the most serious accident to occur during the whole three weeks and the only lost-time one.

Accident 8

A line had to be drained into a drum. The end of the line was 0.5 m above the drum so not surprisingly the operator was splashed on the face with corrosive liquid. Fortunately he was wearing goggles.

A few years earlier there had been a number of accidents during sampling and the need to locate the sample bottle immediately below the sample point had been emphasised in operator training. No-one recognised that the operation being carried out was similar to sampling and that the same precautions should be taken.

Other failures to recognise the true nature of a situation were described in Section 1.2 (no-one realised that a heavy lid under which a man was working produced hazards similar to those of lifting gear) and in Section 7.2 (no-one realised that an open vent is a relief valve).

Accident 9

While three men were lifting a long bent pipe (20 feet by 3 inches bore; 6 m by 75 mm) onto a trailer, one man let go too soon and trapped another man's finger, breaking a bone. He did not lose any time.

Could men be persuaded in cases like this to lift the pipe with straps rather than bare hands?

Accident 10

A 1/4 inch (6 mm) nipple on an instrument line was screwed into a 1/4 inch to 3/8 inch reducing bush (Figure 15.2). The nipple was leaking. While an artificer was trying to tighten it the bush broke and he was splashed in the face with a corrosive liquid. Fortunately he was wearing goggles.

The plant had been built about ten years earlier and the connection was below the current standards which required robust flanged or welded connections between process equipment and the first isolation valve. Screwed connections of the type which broke would not have been permitted anywhere, even on instrument lines, for corrosive materials and only

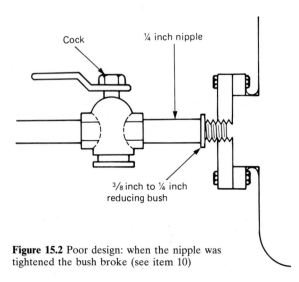

Cock   ¼ inch nipple

³/₈ inch to ¼ inch
reducing bush

Figure 15.2 Poor design: when the nipple was tightened the bush broke (see item 10)

after the first isolation valve for other materials. How far should we go in bringing old plants up to modern standards? Some changes are impossible. Replacement of any connections similar to the one that broke is so cheap and easy that it should certainly be carried out (see Appendix to this chapter). In other cases we have to balance the seriousness of the risk against the cost of improvement.

Perhaps we should list all equipment which is not up to modern standards and decide how far to go in removing it (see Section 22.7).

Accident 11

A laboratory assistant knocked her elbow. She did not know where or when so no recommendations could be made.

Accident 12

A welder lifted up his helmet in order to chip, did not put on ordinary goggles and a piece of slag entered his eye. Similar incidents had occurred before and the welders had been supplied with some helmets in which the blue glass is hinged and there is white glass underneath. When a welder wants to chip he lifts up the blue glass. However, the welders never used these helmets. When asked why they said that the windows were smaller than those in the ordinary helmets. This should not be difficult to put right but people find it easier to fling protective equipment aside as unsuitable rather than suggest improvements.

We noticed that most of the welders found their helmets too long and had cut about 2 inches (50 mm) off them. No-one had ever told the manufacturer.

Accident 13

A fitter burnt his wrist on a small hot, uninsulated pipe. As he stretched his arm he exposed his wrist. The accident could be prevented by a campaign to persuade fitter to wear gloves, but it is not certain that the risk justifies the effort required. Perhaps it is sufficient to publicise the risk and remind fitters that gloves are available. The incident would have been a suitable one for 'tool-box talks' had they been the practice on the works.

Accident 14

A small hydrogen fire occurred in a small open pit in which some residues were dissolved in acid. The flames were lazy, and confined to the pit, and there was no hazard. The fire was soon extinguished with foam. Nevertheless the question was asked, 'Should we continue this operation in the open or would it be safer to change to a closed vessel?'

It was then realised that if this was done the vessel would have to be inerted with nitrogen and fitted with an oxygen alarm; a level measuring instrument would be required and numerous other complications, which were expensive and might go wrong. It was simpler and safer to continue with the present method but a foam generator was permanently installed nearby. I have drawn a line diagram of the equipment that would be required and asked people attending a course to simplify it[4]. Only a few are able to do so.

Accident 15

A craft assistant sprained a finger while helping a fitter to pack a gland on a reciprocating pump. The job was being done by an incorrect but quick method which gives poor results but was often winked at by foremen and engineers. Could a safe and efficient but quicker method have been found? If not, the foremen should have explained why the correct method should have been used and then have enforced its use.

Accident 16

Before moving some liquid from one tank to another an operator failed to check the position of all the valves; a valve leading to a third tank had been left open and this tank was overfilled. Two tonnes of liquid were spilt. The valve had been left open because the line, a long one, was not fitted with thermal relief valves.

It is easy to blame the operator but the mistake was one that it is easy to make. If thermal relief is needed a relief valve should be fitted. We should not in a case such as this rely on operators who may forget to open (or close) a valve.

Accident 17

While making a thermocouple an artificer caught his thumb on the sharp end of the insulation. A tool could be devised so that the operator's thumb is not in direct contact with the insulation but the operation is carried out infrequently and it is therefore doubtful if the effort involved would be justified or if the tool would be found when required.

Accident 18

An electrician was running a temporary cable on a construction site when he stumbled on a piece of wire mesh, lost his balance and hurt his foot on a metal bar.

The foreman told the man to be more careful when crossing ground disturbed by contractors. The engineer endorsed the comment. Our first reaction was to agree. However, when we visited the site we found that the area where the electrician was working had not been disturbed by the contractors. It was covered by weeds, 2 feet tall, which hid numerous tripping hazards, many sharp and sticking up at all angles. The ground should have been bull-dozed by the contractors before work started.

An accident report may seem plausible when read in the office but a visit to the site may tell a different story. An accident report should never be completed before the site has been seen (see Accident 33).

This is the only accident due to untidiness, though in UK industry as a whole it is one of the biggest causes of accidents, the biggest according to some reports.

Accident 19

While picking up some bolts from above eye level a man dislodged some rust which entered his eye. This is another of those eye injuries which could have been prevented only by the wearing of safety spectacles at all times.

Accident 20

A laboratory assistant was wearing thin plastic gloves while handling a corrosive chemical. The gloves split and he burnt his hand. It was agreed that thicker gloves should be used in future.

Accident 21

A choke occurred in a steam-traced line. The insulation was removed and while a man was fixing up a steam lance, to clear the choke, he burnt his hand on the steam tracing. The steam tracing was inadequate and chokes were common. The most effective method of prevention was better tracing, not easy on an old plant but something to note for new plants.

Accident 22

While a process operator was using a wheel-dog to open a valve it moved suddenly and trapped his finger against a bracket. The bracket served no useful purpose and was removed.

Accident 23

Dust in eye. See comments on Accident 19.

Accident 24

A man slipped and fell in an area which was known to be very slippery. The manager commented, 'The need to be more careful will be stressed once again'. The manager, the foreman and the injured man all agreed that nothing could be done and that occasional accidents were inevitable. In fact, it would have been easy to fit a non-slip coating on the wooden blocks on which the man slipped, or to fit studs to his boots. In UK coal mines slipping, a major cause of accidents, has been greatly reduced by fitting anti-slip studs to boots[5].

Accident 25

While climbing into the space below the roof of a floating roof tank, which was being repaired, a foreman bumped his head. His hard hat fell off and he hit his head on the steelwork. Afterwards it was found possible to improve the access.

Accident 26

A man forgot to put on his safety spectacles before using a portable grinder and got a speck of dust in his eye. Though he blamed himself, the oversight is an easy one to make. If he had worn spectacles at all times, the injury would not have occurred.

Accident 27

An electrician caught his wrist on a metal fixing band. Plastic bands are cheaper and safer.

Accident 28

A man was removing empty 40-gallon drums which were stacked four high on their sides (Figure 15.3). The method used was to chock the tier next to the end, remove the end chock and allow four drums to roll down. One drum rolled to one side and as the area was congested the man could not move out of the way. The accident report said, 'To be extra careful when de-stacking drums'. In fact, the only effective method of prevention was to remove the congestion; not easy, as space is limited.

Accident 29

Full 40-gallon drums were being loaded onto a flat wagon with a fork lift truck. The drums were stored on their sides in tiers, as in Figure 15.3. Two drums were lifted onto the wagon at a time (Figure 15.4) and then tipped into a vertical position with the ends of the forks (Figure 15.5). One tipped too far and fell onto the foot of the driver who was standing on the wagon to adjust the position of the drums.

Several similar accidents had occurred before. Despite several investigations no better method had been devised and the problem had been put aside as unsolvable.

Figure 15.3 Method used for storing drums (see Accident 28)

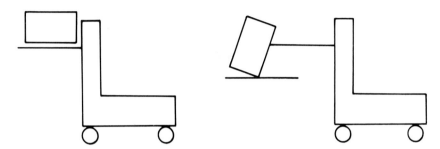

Figure 15.4 Drums are lifted two at a time like this onto a flat wagon

Figure 15.5 ... and then tipped up like this (see Accident 29)

One possible solution would have been to store the drums on their ends and lift them onto the wagons with a suitable attachment fitted to the fork lift truck. However, this would need extra storage space and would not be the complete answer as a layer of horizontal drums was put on top of the vertical layer (Figure 15.6).

One customer would not accept wagons with the upper horizontal layer of drums as a man had been killed when one of these drums fell off. This illustrates one of the problems of industrial safety. There may be a case for forbidding the top layer of drums at all times; there may be a case for allowing it to continue. There is no logic in stopping it at the site where the accident occurred but allowing it to continue elsewhere.

We felt that this problem called for more expertise than we, or any of the local management, possessed and we recommended that it was referred to consultants.

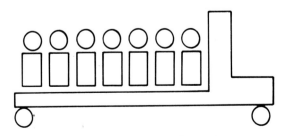

Figure 15.6 Method of carrying drums on a flat wagon (see Accident 29)

Accident 30

A craft assistant cut his finger on a ragged thread while cleaning the threads of a flange with a cloth and cleaning fluid. Safety apart, cloths should not be used for cleaning threads as bits of the cloth get left behind. A brush should be used.

Accident 31

While hoisting a piece of pipe onto a trailer, a speck of dust fell into a man's eye. he was wearing sight-correcting spectacles so safety spectacles would not have helped, unless fitted with side-pieces.

Accident 32

An operator slipped on the deck of a ship (the Works had its own jetty). As with Accident 24, studded footwear might have prevented this accident but in this case we had to balance the decreased risk of slipping with the slightly greater risk that the studs would cause a spark on a concrete floor and ignite any flammable vapour which was present. The latter risk was very slight.

Accident 33

A fitter was fixing a clip onto a hose using a clamping machine. It had been mounted parallel to the edge of the bench instead of at right angles, so that the fitter caught his finger on an object on the bench. The accident report did not say this. As with Accident 18, a visit to the site of the accident told us much more than the report.

Accident 34

A man slipped while entering the Works, put his hand out to steady himself and caught it on a sharp protrusion on the gate. The protrusion, and another one, were removed. The gate is never closed so there is probably no need for it to be there.

Accident 35

Dust in eye, while using an impact wrench out-of-doors on a windy day. See accidents 19 and 23.

Accident 36

A planner slipped while inspecting a construction site. See Accidents 24 and 32.

15.2   General comments

As with the serious accidents discussed in earlier chapters, these trivial accidents can teach us far more than is apparent at first sight. The local managers usually failed to see the inner layers of the onion, partly because they did not realise that there were inner layers waiting to be seen but also because they did not consider that trivial injuries were worth much of their time or attention. As already stated only one accident, involving a motor cycle, caused absence from work. However, many more could have caused more serious injury but by good fortune did not do so.

The most important general lessons to come out of the study were:

(1) If we believe that the accident is so unlikely to happen again that we should do nothing, we should be honest and say so, and not recommend actions that we have no intention of carrying out (see Accident 5). (Of course, after a serious accident we may have to do more than we think is technically necessary, in order to re-assure those exposed.)

(2) A procedure is needed for the control of plant modifications. They should not be made unless authorised by a responsible member of management who should first try to identify the consequences and should also specify the standard to be followed (see Accident 6 and Sections 7.1 and 8.2).

(3) Men should not enter clouds of flammable gas unless protected by water spray (see Accident 6 and Section 4.2).

(4) We should carry out periodic inspections to identify sub-standard equipment (see Accidents 6, 10 and 16) and procedures (see Accidents 4, 8, 15, 28, and 29). These inspections can be carried out in two ways:

(a) A walk round looking for anything sub-standard. It is a good plan to take a camera loaded with slide film with you and photograph the hazards that you see. They can then be shown to those that work in the area. People are shocked to see the hazards that they have passed every day without noticing.

(b) However, hazards are more likely to be found if we look at specific items of equipment such as drain points or screwed joints or specific practices such as sampling or drum handling. Recent incidents, on our own works and elsewhere, and a glance at old reports, will provide subjects for inspection.

(5) Visit the scene of an accident. We will get misleading impressions if we rely solely on written reports (see Accidents 18 and 33).

(6) If protective equipment is not being used, ask why (see Accident 12).

(7) Reports which simply suggest that someone should take more care should be investigated further. There is usually something that managers can do to reduce the opportunities for human error or to protect against the consequences (see Accidents 1, 2, 18, 24, and 28 and reference 6).

(8) Do not rely on procedures when there is a cheap and simple way of removing a hazard by modifying the plant (see Accident 16). Safety by

design should always be our aim. Very often there is no 'reasonably practicable' design solution, as these incidents show, and we have to depend on procedures; but if there is one we should make use of it.

Analysis of the accidents suggests that:

- Twelve accidents, 33% of the total (6, 7, 8, 10, 14, 16, 17, 21, 22, 25, 28, and 34), could have been prevented by better design or layout; the changes required were mostly minor.
- Seven accidents, 20% of the total (4, 5, 9, 15, 29, 30 and 33), could have been prevented by better methods of working.
- Thirteen accidents, 36% of the total (1, 2, 12, 13, 19, 20, 23, 24, 26, 31, 32, 35 and 36), could have been prevented by better protective clothing. The Works handles a lot of corrosive chemicals and the figure might be lower elsewhere. On the other hand this group includes a number of 'dust in eye' incidents which could have been prevented by the wearing of spectacles.
- One accident, 3% of the total (18), could have been prevented by better tidiness.
- Insufficient information was available for three of the accidents, 8% of the total (3, 11 and 27).

(9) People need to be reminded of the limitations of experience. Twenty years without an accident does not prove that the operation is safe unless an accident in the 21st year is acceptable. To be precise, twenty years experience does not even prove that the average accident rate is less than once in twenty years. All it tells us is that we can be 86% confident that the average rate is less than once in ten years (see Accident 4 and reference 3.)

15.3   Conclusions

The incidents described in this chapter have confirmed the theme of this book, that there is much more to be learnt from accidents than we usually learn, not because we are not aware of the facts but because we do not consider them deeply enough. In particular, valuable lessons can be learnt from trivial accidents (see also Chapter 1). In addition, two more general conclusions may be drawn:

(1) Too much writing and talking on safety is concerned with generalities. We can learn much more if we discuss specific incidents. No one should ever generalise without at least describing some incidents which illustrate and support his case (see Part 10 of the Introduction).

(2) To quote Grimaldi[7]:

> (Improvement in safety) is more certain when managers apply the same rigorous and positive administrative persuasiveness that underlies success in any business function ... outstanding safety performances occur when

management does its job well. A low accident rate, like efficient production, is an implicit consequence of managerial control.

A similar view was expressed over twenty years ago following an enquiry into factory accidents by two Factory Inspectors[8]:

> It also produced evidence to confirm that established and generally accepted methods of accident prevention did succeed; and several impressive examples were found of improvements achieved by the energetic and diligent application of principles which had long been advocated, but which had not been put into practice earlier with sufficient thoroughness.

References

1 Ferry, T. S., *Modern Accident Investigation and Analysis*, 2nd edition, Wiley, New York, 1988, Chapter 14.
2 Howell, D., *Health and Safety at Work*, Vol. 12, No. 2, February 1990, p. 26.
3 Kletz, T. A., *Improving Chemical Industry Practices – A New Look at Old Myths of the Chemical Industry*, Hemisphere, New York, 1990, UK, 1984, Section 33.
4 Kletz, T. A., *Plant Design for Safety – A User-Friendly Approach*, Hemisphere, New York, 1991, Section 8.3.2.
5 *Tungsten Carbide Tipped Studs for Anti-Slip Boots*, Safety in Mines Research Establishment, Sheffield, UK, 1971.
6 Kletz, T. A., *An Engineer's View of Human Error*, 2nd edition, Institution of Chemical Engineers, Rugby, UK, 1991.
7 Grimaldi, J. V., *Management and Industrial Safety Achievement*, Information Sheet No 13, International Occupational Health and Safety Information Centre (CIS), Geneva, Switzerland, 1966.
8 *Employment and Productivity Gazette*, October 1968, p. 827.

Appendix: An example of a well-written accident report

Reports such as the following were produced for all the dangerous occurrences, for all the chemical burns and for those other accidents where injuries might easily have been more serious, for example accidents 28, 29 and 32. Handwritten reports were produced for the other accidents. This report deals with accident 10. It is not the original one.

MINOR ACCIDENT REPORT No

Date & time:	17 June 15.00 h.
Place:	E304 still, ground floor.
Injured person:	Mr T. Tiffy, instrument artificer, days.
Injuries:	Chemical burns to face.
Date of investigation:	18 June

Investigation team: Mr B. Boss, plant manager
                                        Mr T. Trip, process foreman
                                        Mr S. Spanner, maintenance foreman
                                        Mr W. Watchman, safety representative
                                        Mr G. Guardsman, safety officer
                                        Mr P. Pump, process operator (witness)
                                        Mr T. Tiffy

What happened? At about 14.00 hours on 17 June an operator told Mr
Trip that there was a slight leak from a screwed joint on the base of E304
still. After inspecting the leak Mr Trip asked Mr Spanner to repair it and
issued a permit-to-work for the job. It stated that the leaking equipment
contained corrosive materials (-----------) under a hydrostatic pressure of
about 4.5 m (7 psi or 0.5 bar) and that gloves and goggles should be worn.
The permit was produced and was found to be in order. Mr Spanner
accepted the permit and asked Mr Tiffy, an experienced craftsmen, to
tighten the joint. He showed him the permit.

The joint is shown in the attached diagram (see Figure 14.2.) The leak
was coming from the screwed joint between the 1/4 inch (6 mm) nipple
and the reducing bush. The liquid was trickling out, not spraying. Mr Tiffy
attempted to screw the nipple further onto the bush but while he was
doing so the bush broke and a stream of liquid came out. Mr Tiffy was
standing to one side, having foreseen the possibility of the bush breaking,
and the main stream of liquid passed by him but some spray came in
contact with his face. The joint was at chest height. Mr Tiffy was wearing
goggles.

Mr Tiffy immediately put his face under a shower which was located
about 3 m away. He was assisted by Mr Pump, process operator, who
happened to be passing. Mr Pump sent for the ambulance and Mr Tiffy
was taken to the Medical Centre but was not detained. There is some
discoloration on his face which is expected to last for a few days.

Discussion. The joint which broke was installed about 10 years ago when
the plant was built. It is not up to the current standard which states that
screwed connections may be used only on narrow bore instrument lines,
only after the first isolation valves, and only if the equipment contains
non-hazardous materials (that is, does not contain corrosive liquids, toxic
liquids or gases or flashing flammable liquids). (See Engineering Dept
Standard No -.)

While it is clearly impossible to bring all old equipment up to modern
standards the joint which leaked is so far below them that any other
similar equipment should be replaced as soon as possible.

Actions. Mr Boss will arrange for one of the process foreman to be
released from his normal duties for a day, or longer if necessary, so that
he can survey the plant and list any other sub-standard joints. These will

be replaced during the shut-down scheduled for September.

ACTION: Mr B. Boss

Mr S. Spanner

The safety officer will ask other plants that handle corrosive or other hazardous materials to carry out similar surveys.

ACTION: Mr G. Guardsman

The Company Safety Adviser will be asked to inform other factories.

ACTION: Company Safety Adviser

Completion date. See above. This report should be brought forward on 1 September to confirm that sub-standard joints have been listed and are on the shut-down list and again after the shutdown to confirm that they have been replaced.

ACTION: Works office

Resources required. This depends on the number of sub-standard joints found. A quick look suggests that there could easily be 10–20, perhaps more. If so the cost of replacing them will be about £---. This will not affect the length of the shutdown.

B. Boss

Plant manager

Circulation

Works manager	Works engineer
Plant engineer	Those present at enquiry
Engineering Department	Company safety officer
Plant notice board	Works office
Medical officer	
Personnel officer (for file of injured man)	
Managers of plants with similar hazards	

Note added by works manager

There must be many other items of equipment on the works which are below current standards. How far should we go in replacing them? I am writing to the Company Safety Adviser asking him to review this question, with Engineering Department, and prepare a note which will list the types of equipment involved and the extent of the problem and make recommendations.

What would have happened if Mr Pump had not been passing? In some factories operation of a shower sounds an alarm in the control room. Should we adopt this system? What will it cost? Will the safety officer please consider and make recommendations.

I shall also ask the Company Safety Adviser to send a copy of this report, suitably edited, to the Institution of Chemical Engineers Loss Prevention Bulletin, where it can be published anonymously.

B. Bigshot

Chapter 16

Pipe failures

Since piping systems failures are one of the major causes of catastrophes in the oil and chemical industries, the importance of proper installation, maintenance, and operation of process piping systems is essential in preventing major loss of life, property, and income.

B. T. Matusz and D. L. Sadler[1]

This chapter discusses a number of pipe failures. The recommendations made for preventing further failures are important in themselves and also show that by looking at a group of related accidents we may be able to draw conclusions which we could not draw from single incidents.

16.1 Why are pipe failures important?

Since Flixborough (Chapter 8) there has been an explosion of papers on the probability of leaks and on the behaviour of the leaking material – how it disperses in the atmosphere, what pressure is developed if it ignites and so on – and expensive experiments have been carried out. In contrast much less attention has been paid to the reasons why leaks occur and to ways of preventing them. It is as if leaks were inevitable. Yet if we could prevent leaks we would not need to worry so much about their behaviour.

Table 16.1, derived from a review of sixty-seven major leaks of flammable gas or vapour,[2] most but not all of which ignited, shows that half were the results of failures of pipes or pipe fittings, over half if we ignore transport accidents. If we want to prevent major leaks, the most effective action we can take is to prevent pipe failures. Other studies have come to similar conclusions. According to Pitblado *et al.* 35% of leaks are due to failures of pipework (excluding valves) and 25% to failures of valves[3]. The Institution of Chemical Engineers' training module on piping[4] says that 40% of losses are due to pipework failures and an insurance survey showed that ten out nineteen major refinery incidents were due to pipework failures[5].

Table 16.1 Origin of major leaks of flammable gases and vapours

Origin	Number of incidents	Notes
Transport container	10	Includes 1 zeppelin
Pipeline (inc. valve, flange, glass bellows, sight-glass, etc.)	34	Includes 1 sight-glass and 2 flexes
Pump	2	
Vessel	5	Includes 1 internal explosion, 1 foam-over and 1 case of over-heating
Relief valve or vent	8	
Drain valve	4	
Maintenance error	2	
Unknown	2	
Total	67	

There is not sufficient detail in reference 1 to analyse the reasons for the thirty-four pipe failures so instead I have analysed about fifty pipe failures (or near-failures) on which I have seen reports, most of them unpublished. Many of the failures did not result in fires or explosions as the leaks failed to ignite, the contents were non-flammable or the leak was prevented at the last minute.

The incidents are summarised below.

16.2   Some pipe failures that could have been prevented by better design

In some cases the piping designer was not provided with the information needed for adequate design.

(A1) Water was added to an oil stream using a simple T-junction (Figure 16.1). The water did not mix with the oil and extensive corrosion occurred downstream of the junction. There was a major fire.

The corrosion would not have occurred if the water pressure had been higher, so that the water hit the opposite wall of the pipe, or if the water had been added to the centre of the oil line as shown in Figure 16.2.

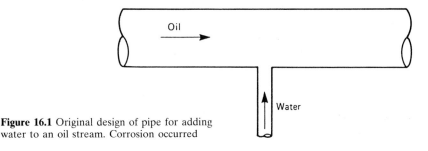

Figure 16.1 Original design of pipe for adding water to an oil stream. Corrosion occurred

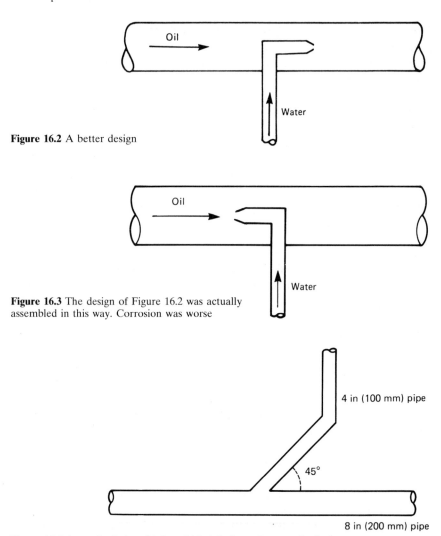

Figure 16.2 A better design

Figure 16.3 The design of Figure 16.2 was actually assembled in this way. Corrosion was worse

Figure 16.4 A poorly designed joint which failed, causing a major leak

(A2) A device to achieve this was designed for another plant. It was assembled as shown in Figure 16.3. Corrosion was worse! Once the device was assembled it was impossible to tell if it had been assembled correctly. It should have been designed so that there was a tell-tale sign on it or, better still, so that it could not be assembled incorrectly.

(A3) A 4-inch (100 mm) branch was fitted onto an 8-inch (200 mm) pipe at an angle of 45° (Figure 16.4). The line failed and 30 tonnes of hot, flammable hydrocarbon were released; fortunately it did not ignite[1].

Failure was the result of vibration and fatigue, enhanced by the poor detailed design of the joint, the different wall thicknesses of the two pipes and the oval shape of the hole. In addition the smaller line should have been better supported and when vibration occurred, the reason for it should have been investigated.

(A4) A valve was opened in error and cryogenic liquid came into contact with a mild steel pipe which disintegrated.

We would never tolerate a situation in which operation of a valve at the wrong time could result in equipment being overpressured. We would install a relief valve. Similarly, we should not allow equipment to be overcooled by operation of a valve at the wrong time. The designer should have specified a grade of steel suitable for low temperature or installed an interlock to prevent the valve being opened if the temperature is low.

This incident illustrates the very different attitudes adopted towards pressure and temperature. Too little or too much of either may be hazardous and, if so, it is equally important to prevent either condition occurring[6].

(A5) Some wet hydrocarbon gases had been blown down for a prolonged period. An ice-hydrate plug blocked an 18-inch (0.46 m) blow-down line. It was cleared by external steaming. When the plug became loose the pressure above it caused it to move with such force that the line fractured at a T-junction.

The process designers did not foresee that wet gases might have to be blown down for so long. It was unlikely to occur very often. Nevertheless the incident would not have occurred if the design team had foreseen that prolonged blowdown might occur and had tried to prevent choking, for example by steam heating.

(A6) A stainless steel line intended for use with hydrogen at 360°C was fitted with a branch leading to a relief valve. There would normally be no flow through the branch and it was calculated that 1 m along it the temperature would have fallen below 100°C and that mild steel could be used. However, the line failed by hydrogen attack. Possibly the method used for calculating the temperature in the line did not allow for the shielding effect of surrounding pipes.

(A7) A bellows was found to be distorted. It had been designed for normal operation but the piping designer had not been told that it would be hotter when it was steamed out before a shutdown.

(A8) A line carrying heat transfer oil failed by fatigue as the result of repeated expansion and contraction as the temperature varied. The oil fell onto some cables and attacked them and a short circuit ignited the oil. There should have been more expansion bends in the line. (In addition, flanges on oil lines should be located so that any drips do not fall on cables.)

(A9) Water froze in an LPG drain line and fractured a screwed joint. Screwed joints are not suitable for process materials except for small bore lines after the first isolation valve (see Chapter 15, Accident 10). Flanged or welded joints should be specified.

(A10) A valve in a 10-inch (250 mm) liquefied butane line was located in a pit which was full of rainwater contaminated by sulphuric acid from a leaking underground line nearby. The bolts holding down the valve bonnet corroded and the bonnet flew off. A massive leak of butane exploded, killing seven people and causing extensive damage.

At first there was only a small leak from the valve. It was decided to empty the line by washing it out with water at a gauge pressure of 110 psi (7 bar), more than the usual operating pressure (50 psi or 3.3 bar). The line was designed to withstand this pressure but in its corroded state it could not do so[7].

Although it was not good practice to install a valve in a pit that was liable to be flooded it was also not good practice for the operating team to tolerate the flooding or to try to sweep out a corroded line using water at a higher pressure than the pressure normally used.

(A11) A beryllium/copper circlip was used to secure the joints in an articulated arm carrying ammonia. The joint blew wide open. Stainless steel should have been specified. Afterwards it was decided to stock only stainless steel circlips so that errors cannot occur.

(A12) The feed line to an ammonia plant failed suddenly as the result of hydrogen attack. The grade of steel used was considered suitable for the duty at the time the plant was built (about 1970) but since then the Nelson curves, which define the conditions under which hydrogen attack will occur, have been revised[8]. This is the only pipe failure I know of which occurred as a result of an error in a code.

(A13) On a methanol plant, the joint between a 6-inch (150 mm) synthesis gas pipe and a flange failed and the escaping gas exploded. No one was hurt but damage was extensive. The report said that weld-neck flanges should be used for cyclic operations as they have ten times the fatigue life of the joint actually used, a combination of lap-joint and stub-end[9].

16.3   Some pipe failures that could have been prevented by better inspection during or after construction

(B1) A temporary support across an expansion loop was left in position. Fortunately it was spotted as the plant was warming up.

(B2) The exit pipe from a high pressure converter was made from mild steel instead of 0.5% Mo. The pipe failed by hydrogen attack, a leak occurred at a bend and the reaction force pushed the converter over.

This was the most spectacular of many incidents that have occurred because the wrong material of construction was used. Whenever the use of the correct material is critical this should be indicated on the drawings and all incoming steelwork (pipes, flanges, welding rods as well as fabricated items) should be checked before installation. Every piece of pipe, flange, etc., should be checked, not just a sample[10,11].

Reduction in the number of grades of steel used will reduce the chances of error. It is false economy to specify a cheaper grade for a limited

number of applications if it increases the chance of error, during maintenance as well as construction (see item A11).

Six pipes in an ammonia plant were found to be cracked. The correct
grade of steel had been used but it had received the wrong heat treatment[12].

(B3) A construction worker cut a hole in a pipe-line at the wrong place.
Discovering his error, he welded on a patch but said nothing. The line was
radiographed but this weld was omitted as the radiographer did not know
it was there. The line was then insulated.

The weld was sub-standard, leaked and several men were gassed by
phosgene, one seriously.

It is easy to talk of criminal negligence but did the man understand the
importance of good welding, realise the nature of the materials that would
be in the pipeline or what might happen to other people if his work was
sub-standard? Unfortunately, most construction engineers do not believe
it is practicable to try to explain these things to construction workers.

(B4) Underground propane and oxygen lines leaked and an explosion
occurred underground. The report said, 'During the construction of the
pipework, doubts were expressed by the works management as to the
quality of the workmanship and the qualifications of those workers
employed'.

Why did they not do more than just express doubts to each other?

It is not good practice to run pipes underground in a factory as the
ground is often contaminated by chemicals which cause corrosion
(although that did not occur in this case). If pipes, however, are run underground they should be wrapped, surrounded by clean sand or gravel and
cathodically protected. When an unprotected line leaked, 140 tonnes of
product were lost over a period of four days through a hole only ⅛ to ¼
inch (3–6 mm) diameter. As the leak occurred over a public holiday the
records office did not pick up the discrepancy between the quantities
despatched and received.

A leak in an underground line contributed to incident A10.

(B5) A portable hand-held compressed air grinder, in use on a new
pipeline, was left resting on a liquefied petroleum gas line. The grinder
was not switched off when the air compressor was shut down. When the
air compressor was started up some of the line was ground away.

(B6) Several hangers failed and a new pipe sagged, over a length of
14 m. When the pipe was installed, it did not have the required slope, so
the contractors cut some of the hangers and welded them together again.
One of the welds failed. Other hangers failed as a result of incorrect
assembly and lack of lubrication.

After a 22-inch (0.56 m) low pressure steam main developed a crack it
was found that the spring on one support was fully compressed, that a
support shown on the drawing was not fitted, that another support was
fitted but not attached to the pipe and that the nuts on another support
were slack.

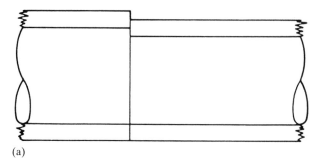

(a)

(b)

Figure 16.5 (a) and (b) Two pipes of slightly different diameter were welded together

(B7) Two lengths of 8-inch (200 mm) pipe which had to be welded together were not exactly the same diameter so the welder joined them with a step between them over part of the circumference (Figure 16.5(a) and (b)). The pipe was then insulated. The botched job was not discovered until ten years later when the insulation was removed for inspection of the welds.

(B8) A bellows blew apart a few hours after installation. The split rings which supported the convolutions and equalised expansion were slack, with gaps of up to 1/2 inch (13 mm) between the butts of the half rings. The plant was not inspected before commissioning.

On another occasion a bellows was damaged before delivery and this was not noticed by the construction team.

Other bellows have failed because they were installed between two pipes which were not exactly in line and the bellows were allowed to take up the misalignment. If fixed piping does not meet exactly it can be pushed until it does. If bellows are used the piping must be made with more, not

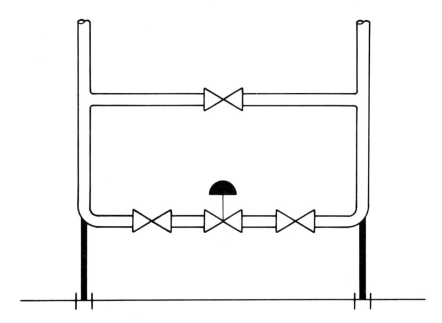

Figure 16.6 The supports were bolted to the concrete bases and wleded to the pipes. When the pipe moved, a piece was torn out

less, accuracy. Bellows should not be used to compensate for misalignment, unless specially designed for that purpose.

A near failure of a bellows was described in item A7. The most famous bellows failure of all time – Flixborough – was described in Chapter 8 where it was suggested that when hazardous materials are being handled designers should avoid the use of bellows and use expansion loops instead. What you don't install, cannot be badly installed.

(B9) The supports shown in Figure 16.6 were bolted to the concrete bases and welded to the pipes. When the control valve closed suddenly, the shock caused the pipe to move suddenly and a piece was torn out of one of the bends. The escaping liquid caught fire and damage was extensive[13].

It is unlikely that the designer asked for such rigid fixing of the pipe. The details were probably left to the construction team.

(B10) A 10-inch (250 mm) pipe was fitted with a ¾ inch (20 mm) branch. The main pipe rested on a girder and there was a gap of 5 inches (125 mm) between the branch and the girder (Figure 16.7). When the pipe was brought into use its temperature rose to 300°C. The branch came into contact with the girder and was knocked off. Calculation showed that the pipe, which was 120 feet (36 m) long, had moved 6 inches (150 mm).

Again, it is unlikely that the designer specified precisely how the branch was to be positioned in relation to the girder and that he left this detail to be decided on site. The construction workers would not have known the operating temperature of the pipe or how much it would expand.

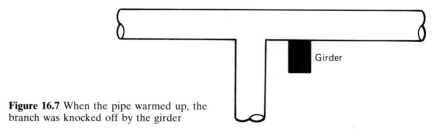

Figure 16.7 When the pipe warmed up, the branch was knocked off by the girder

(B11) An old pipe was re-used. Its condition was not checked and it failed as a result of corrosion/erosion which had occurred on the old duty.

On another occasion an old pipe that was re-used had used up much of its creep life and failed in service though a new pipe would not have failed. A stream of high pressure gas produced a flame 30 m long. The pipe that failed had been used for 12 years at 500°C but creep cannot be detected until a pipe is close to bursting.

(B12) A pipe was laid on the ground and corroded. The designer probably left the choice of support to the construction team.

(B13) Several pipes were inadequately supported, vibrated and failed by fatigue. Supports for small bore pipes are often not specified by designers but decided on site. It is often difficult to know whether or not a pipe will vibrate until the plant is on line and perhaps therefore the most effective way of preventing these incidents is for the operating team to take action when pipes are vibrating. Unfortunately they are often too busy during start-up and then the pipes become part of the scene and are not noticed.

(B14) A line carrying a material of 40°C melting point was kept hot by steam tracing. One section of tracing was isolated and the line set solid. Expansion of liquid in the rest of the line caused it to burst (Figure 16.8).

The detail of the steam tracing was probably not specified by the designer but left to the construction team. It should not have been possible to isolate the heating on a part of the line.

(B15) A level controller fractured at a weld, producing a massive release of oil, hydrogen and hydrogen sulphide. The report said that the

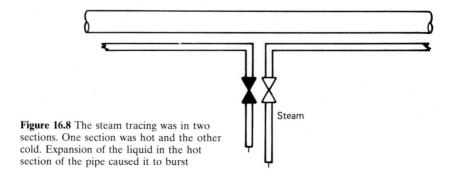

Figure 16.8 The steam tracing was in two sections. One section was hot and the other cold. Expansion of the liquid in the hot section of the pipe caused it to burst

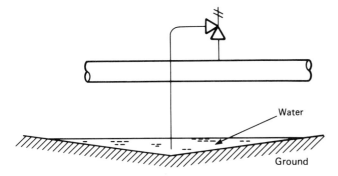

Figure 16.9 The end of the relief valve tail pipe was so close to the ground that it was below the surface of a puddle and was blocked when the puddle froze

failure was due to poor workmanship and 'emphasises the need for clear instructions on all drawings and adequate inspection during manufacture.'

(B16) A 1-inch (25 mm) screwed nipple blew out of a line carrying heavy oil at 350°C. Most of the plant was covered by an oil mist 30 m deep. Fortunately it did not ignite.

The nipple had been installed twenty years earlier, during construction, for pressure testing and was not shown on any drawing. If its existence had been known it would have been replaced by a welded plug.

(B17) A line carrying liquefied gas was protected by a small relief valve which discharged onto the ground. The ground was not level and after heavy rain the end of the tailpipe was below the surface of a puddle (Figure 16.9). The water froze and the line was overpressured.

(B18) Sometimes it is necessary to reinforce the wall of a pipe, where there are additional stresses from a branch or a support, by welding on a plate. There is usually a small gap, up to 1/16 inch (1.6 mm), between the plate and the wall of the pipe. This space should be vented, by a 1/4 inch (6 mm) hole or a gap in the welding, or a pressure may develop when the pipe gets hot. The rise in pressure will be greater if any water is trapped in the space and turns to steam.

A steam pipe designed to operate at a gauge pressure of 200 psi (14 bar) collapsed because a reinforcing pad was not vented. A blowdown main collapsed for the same reason, fortunately during heat treatment.

(B19) Many pipes have failed because water collected in dead-ends and froze, or corrosive materials dissolved in the water and corroded the pipe. Most oil streams contain traces of water. A 3-m branch on a natural gas pipeline was corroded in this way; the escaping gas ignited and three men were killed[14].

Dead-ends are often the result of modifications but they are sometimes installed in new plants to make extension easier. Dead-ends should point upwards, not downwards, so that water cannot collect in them. If the

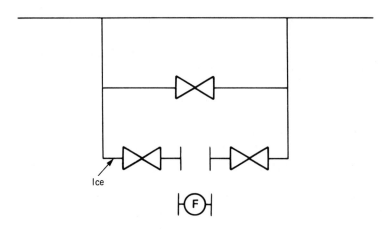

Figure 16.10 When the flowmeter was removed it formed a dead-end in which water collected

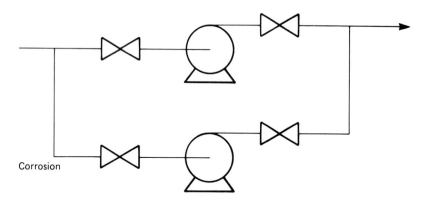

Figure 16.11 The lower pump was rarely used. Water collected and corrosion occurred

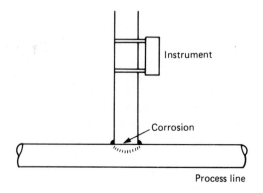

Figure 16.12 Water collected in a piece of pipe, used as an instrument support, and corrosion occurred

liquid in the pipe is denser than water then dead-ends should point downwards.

Dead-ends may be formed by removal of unused equipment (Figure 16.10) or simply by not using equipment. One of the two pumps shown in Figure 16.11 was rarely used. It was at a slightly lower height than the other pump so that water collected in the branch which corroded right through. Prevention of incidents such as these lies more in the hands of the operating team than the constructors.

A piece of pipe was welded onto a liquefied petroleum gas line to support an instrument (Figure 16.12). Water collected in the support and corroded the LPG line. Gas was seen blowing out of the support!

(B20) A leak occurred on a T-piece on a high pressure boiler feed water main four years after start-up. The construction team had found that insufficient forged T-pieces were available and had therefore made three by welding together short lengths of pipe. If they had told the operating team that they had done this then the sub-standard T-pieces could have been scheduled for replacement at the first opportunity. Keeping quiet about the bodging compounded the original error.

(B21) As this section has been a catalogue of construction failures, and of failure to spot the failures, let us end it with a success story. An alert welding inspector was looking through the radiographs of some pipework intended for use with refrigerated ammonia. He noticed that a series of radiographs said to have been taken from different butt welds were in fact actually taken on the same weld. The inspector had been engaged by the client, not the contractor[15].

16.4   Some pipe failures that could have been prevented by better operations

Some such failures have already been mentioned. See items A10, B13 and B19. Here are a few more.

(C1) A crane was used to move a line full of highly flammable liquid so that a joint could be made. A branch on the line was knocked off and a leak occurred.

(C2) After completing a lift a crane had to be moved about 100 m along a road for its next job. The driver decided not to bother lowering the job and he hit and damaged an overhead pipeline.

If cranes frequently travel along a roadway, any pipes that cross can be protected by 'goalposts', girders across the road a few metres either side of the pipes.

(C3) An unforeseen decomposition occurred in a reactor, the temperature in the exit pipe rose rapidly and the pipe ruptured.

(C4) A little-used line was left full of water during the winter and was split by frost.

16.5   Conclusions

Some recommendations that arise out of a single incident or a few incidents have already been made. In this section we try to draw some conclusions from the incidents as a whole.

The classification of some of the incidents may be disputed but the general conclusion seems inescapable: The most effective action we can take to prevent pipe failures is to:

- Specify designs in detail.
- Inspect thoroughly during and after construction to make sure that the design has been followed and that good engineering practice has been followed when the design has not been specified in detail but left to the discretion of the construction team. Much more thorough inspection is needed than has been customary in the past.
- Perhaps try to explain to construction workers why it is important to follow the design in detail and to construct details not specified in the design in accordance with good engineering practice.

In the UK the Health and Safety Executive (HSE) have made regulations for the inspection and certification of pressure systems[16] but they emphasise ongoing inspection throughout the life of the plant. Important as this is, the incidents described above suggest that inspection during and after construction will be much more effective in preventing failure.

On cross-country pipelines the HSE require a quantitative assessment of the risk and they have described the methods they use[17,18].

Who should carry out inspections during and after construction? Construction teams usually employ their own inspectors but they often are (or used to be) fairly junior people who do not carry much weight and often do not know what to look for. If radiographs are specified they see that they are done but they may not have the knowledge to detect the departures from good engineering practice that are listed below. I suggest that inspections are carried out by:

- A member of the design team – he will notice departures from the design intention – and
- a member of the start-up team – he will have a greater incentive than anyone as he will suffer from the results of the construction defects.

The following is a list of some of the things that should be looked for during these inspections. The numbers refer to the descriptions of the incidents.

During construction
- Equipment has been made of the grade of steel specified and has received the specified heat treatment (A11, B2).
- Old pipe is not re-used without checking that it is suitable for the new duty (B11).

- Pipes are not laid underground (B4).
- Workmanship is of the quality specified and tests are carried out as specified (B3, B4, B15).
- Purchased equipment is undamaged (B8).

After construction and before start-up
- Pipes are not secured too rigidly (B9).
- Pipes are free to expand and will not foul supports or other fixtures when they expand (A8, B10).
- Flanges on liquid lines are not located above cables (A8).
- Supports are correctly installed and assembled and springs are not fully compressed or extended (B6).
- There are no obviously sub-standard joints (A3).
- Pipes are not touching the ground (B12).
- Temporary supports have been removed (B1).
- Temporary branches, nipples and plugs have been removed and replaced by properly designed welded plugs (B16).
- Screwed joints have not been used (A9).
- Trace heating cannot be isolated on part of a line without isolating the whole (B14).
- Equipment has not been assembled wrongly. First identify any equipment which can be assembled wrongly (A2).
- Pipes do not pass through pits or depressions which can fill with water (A10).
- Relief valve tailpipes or drain lines are not so close to the ground that they may be blocked by ice or dirt (B17).
- Lines which contain water can be drained when not in use (C4).
- The slope of lines is correct, for example, blowdown lines should slope towards the blowdown drum (B6).
- There is no bodging (B6, B7, B20).
- Reinforcement pads are vented (B18).
- There are no dead-ends in which water or corrosive materials can collect (B19). Note that dead-ends include little-used branches as well as blanked branches.
- There are no water traps in which water can collect (B19).
- Bellows are not distorted and any support rings are not loose (B8).

After start-up
- Pipes are not vibrating (B13).

Note that the incidents do not suggest that there is anything wrong with the piping codes. (Item A12 described a rare incident due an error in a code.) We do not need stronger, thicker pipes. The failures occurred because the codes were not followed or because the plant was not constructed in accordance with good engineering practice. This is a rather elusive quality that is not usually written down. An attempt to list a few of the factors that contribute towards it has been made above. The list is not intended to be complete. It merely summarises the points that come out of the incidents described. It is intended as a starting point to which readers can add their own experiences. The construction inspector has to look out for things that no one has ever dreamt of specifically prohibiting.

As already stated, other writers agree that pipework is the site of most leaks. When it comes to the causes of the leaks another review is broadly in agreement with my conclusions but attaches more importance to operating errors[19].

Hurst and co-workers[20] have analysed pipe failures under three headings: the direct (or immediate) cause, such as pressure or vibration; the underlying cause, such as construction or maintenance; and the mechanism which might have prevented the failure, such as hazard studies, checks or human factors reviews. In contrast to the conclusions of this chapter they conclude that construction was the underlying cause in only about 10% of the failures and that the most important underlying causes were maintenance (nearly 40%) and design (27%). They recommend more human factor reviews and hazard analyses of maintenance and process tasks and equipment designs and routine inspections but make no special mention of inspection after construction. Details of the pipe failures are not given so readers cannot see if they agree with the allocations.

The objective of Hurst's work is to see if it is possible to multiply the published failure rates of pipes (and vessels) by a factor which measures the competence of the management[21].

Finally, if we can reduce the quantity of hazardous material in the plant, or use a safer material instead, as discussed in Chapters 8 and 9, we need worry less about pipe failures.

References

This chapter is based on a paper which was published in *Plant/Operations Progress*, Vol. 3, No. 1, Jan. 1984, p. 19, and thanks are due to the American Institute of Chemical Engineers for permission to quote from it.

1 Matusz, B. T. and Sadler, D. L., 'A comprehensive program for preventing cyclohexane oxidation process piping failures', *Proceedings of the American Institute of Chemical Engineers Loss Prevention Symposium*, Houston, Texas, March 1993.

2 Davenport, J.A. *Chemical Engineering Progress*, Vol. 73, No. 9, Sept. 1977, p. 54.

3 Pitblado, R. M., Williams, J. C. and Salter, D. H., *Plant/Operations Progress*, Vol. 9, No. 3, July 1990, p. 169.

4 *Safer Piping*, Hazard Workshop Module No 12, Institution of Chemical Engineers, Rugby, UK, undated.

5 *Loss Prevention Bulletin*, No. 99, June 1991, p. 1.

6 Kletz, T. A., *Improving Chemical Industry Practices – A New Look at Old Myths of the Chemical Industry*, Hemisphere, New York, 1990, Section 2.

7 Vervalin, C. H., *Fire Protection Handbook for Hydrocarbon Processing Plants*, Vol. 1, 3rd edition, 1985, p. 122.

8 Prescott, G. R. Braun, C.F., Blommoert, P. and Grisolia, L., *Plant/Operations Progress*, Vol. 5, No. 3, July 1986, p. 155.
9 Lloyd, W. D., *Plant/Operations Progress*, Vol. 2, No. 2, April 1983, p. 120.
10 Vincent, G. C. and Gent, C. W., *Ammonia Plant Safety*, Vol. 20, 1978, p. 22.
11 Dunmore, O. J. and Duff, J. C., *Materials Performance*, July 1974, p. 25.
12 Lawrence, G. M., *Plant/Operations Progress*, Vol. 5, No. 3, July 1986, p. 175.
13 Geisler, V. G., *Loss Prevention*, Vol. 12, 1979, p. 9.
14 *Safety Recommendations Nos P-75-14 & 15*, US National Transportation Safety Board, Washington, DC, 14 November 1975.
15 Health and Safety Executive, *Manufacturing and Service Industries 1982*, Her Majesty's Stationery Office, London, 1983, p. 15.
16 Health and Safety Executive, *Pressure Systems and Transportable Gas Containers Regulations 1989*, Her Majesty's Stationery Office, London, 1989. See also *Safety of Pressure Systems: Approved Code of Practice*, 1990.
17 Jones, D. A. and Gye, T., 'Pipelines protection – How protective measures can influence a risk assessment of pipelines', *Pipelines Protection Conference*, Cannes, France, 1991.
18 Carter, D. A., *Journal of Loss Prevention in the Process Industries*, Vol. 4, No. 1, Jan. 1991, p. 68.
19 Blything, K. W. and Parry, S. T., *Pipework Failures – A Review of Historical Incidents*, Report No SRD R441, UK Atomic Energy Authority, Warrington, UK, 1988.
20 Geyer, T. A. W., Bellamy, L.J., Astley, J.P. and Hurst, N.W., *Chemical Engineering Progress*, Vol. 86, No. 11, Nov. 1990, p. 66.
21 Hurst N. W., Bellamy, L. J. and Wright, M. S., 'Research models of safety management of onshore major hazards and their possible application to to offshore safety' in *Major Hazards Onshore and Offshore*, Symposium Series No. 130, Institution of Chemical Engineers, Rugby, UK, 1992, p. 129.

Chapter 17

Piper Alpha

Brian Appleton

Behold, how great a matter a little fire kindles.

Epistle of St James, 3:5

The destruction of the Piper Alpha offshore oil platform in 1988 had a similar effect on the UK offshore oil and gas industry as Flixborough had on the chemical industry in the UK in the late 1970s. The destruction of the platform by an explosion and the subsequent massive oil and gas fire and the death of 167 men showed that the potential hazards of the offshore industry were greater than had been generally believed by the public and perhaps by the industry itself. The government set up a public inquiry and its findings and recommendations[1] led to a radical change in the approach to offshore hazards demanded of managements and of the regulatory authorities.

Piper Alpha was a production platform located in the North Sea about 180 km north-east of Aberdeen. The platform had facilities to drill wells to the producing reservoir and extract, separate and process the reservoir fluids, a mixture of oil, gas and water. Gas and water were separated from the oil in production separators. Gas condensate liquid, almost entirely propane, was separated from the gas by cooling and was then reinjected back into the oil to be transported to shore by pipeline. Gas was also exported to shore via a gas gathering platform 55 km from Piper. Piper Alpha not only exported its own production to shore but was also linked to two other platforms, Tartan, 20 km, and Claymore, 35 km away, by gas pipelines, and the oil production from both these platforms was linked into the Piper oil export line to shore.

The disaster originated at 20.00 hours on 6 July 1988 with the explosion of a low lying cloud of condensate. That caused extensive damage and led to an immediate large crude oil fire, the smoke from which made access to any of the platform's lifeboats impossible from the outset. About 20 minutes later there was a second explosion which caused a massive intensification of the fire. It was due to the rupture of the riser on the gas pipeline from the Tartan platform. The fire was further intensified by

Figure 17.1 The fire on Piper Alpha during its early stages, before the rupture of the risers. (Copyright: Robert Gibson)

successive ruptures some 30 minutes and 60 minutes later of risers on the other two gas pipelines, to the gas gathering and the Claymore platforms. Apart from the drilling rig end structure, the Piper platform was totally destroyed within a few hours, its equipment and accommodation falling into the sea which was about 150 m deep (Figure 17.1).

The main loss of life took place in the accommodation block. Most of the men onboard were off duty or in bed and many of those working the night shift made their way there immediately after the initial explosion. The normal method of transport to and from the platforms was by helicopter and with the lifeboats inaccessible the crew expected to be rescued by helicopter. However no helicopter could land in the face of the oil smoke and massive flames and most of those who died did so from carbon monoxide poisoning due to smoke saturating the accommodation building. Some men did escape by jumping from the platform into the sea, which luckily was calm that evening, and they were picked up by fast rescue craft launched from ships in the vicinity.

The exhaustive public inquiry not only brought out the detailed events of the disaster but also examined the background to some of the deficiencies that occurred on the fatal evening. This chapter sets out six major factors which contributed to the accident and high loss of life and develops from these some fundamental lessons about managing safety in a large-scale process industry.

17.1   The initial explosion

The leak of condensate which ignited to cause the initial explosion originated in the process night shift's response to the tripping out of the pump which reinjected condensate into the export oil line. After trying in vain to restart the tripped pump the operators decided to commission the installed spare pump. They knew that a permit-to-work (PTW) for that pump had been issued earlier in the day with the intention of carrying out a major overhaul lasting about two weeks. They were also aware that, while the pump motor had been electrically isolated, its suction and delivery valves closed and the pump drained down, none of the equipment had been opened up and the lines around the pump had not been slip-plated off. Accordingly it would be simple and quick to reconnect the power supply and restart the pump.

However, they were not aware, but should have been, that a second maintenance job had been started on that spare pump during the day. A relief valve on its delivery pipe had been taken off in order to check its set pressure. The work had not been completed by 18.00 hours, the end of the day shift, and refitting the relief valve had been left over for the following day. When the night shift opened up the pump's suction valve in preparation for starting-up, condensate leaked from the site of the removed relief valve. A blank flange assembly at the site of the relief valve was not leak tight. It was that leak which ignited and exploded (Figure 17.2).

The lack of awareness of the removal of the relief valve resulted from failures in the communication of information at shift handover at 1800 hours and in the operation of the PTW system in relation to its removal. Three handovers, those between lead process operators, the condensate area operators, and the maintenance lead operators should have, but did not, include communication of the fact that the relief valve had been removed and had not yet been replaced. However, what is passed on at a shift change cannot be defined exactly and has to be left to the judgement of those involved. That is one reason why throughout the process industry a PTW system is employed. Such a system is a formal, documented method used to control and pass information about maintenance work. It ensures that process equipment is properly isolated and vented before it is opened up, that isolation is maintained throughout the duration of the work, that the correct safety measures are used by maintenance personnel, and that the reinstated equipment is checked before it is returned to service.

On Piper Alpha there were deficiencies in the operation of the PTW system with regard to the relief valve work. The work was carried out by a contract maintenance organisation whose supervisor did not inspect the job site before suspending the permit overnight. He did not discuss the non-completion of the work with the process supervisor but signed off the permit and left it on the control room desk. The condensate injection pump suction and delivery valves were not secured against inadvertent

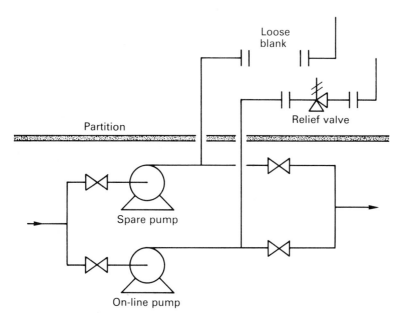

Figure 17.2 Piper Alpha: the on-line pump broke down so the other pump, which was shut down for overhaul and for testing of the relief valve, was brought on line. The operators did not know that the relief valve had already been removed and the open end loosely blanked

opening by locks. The permits for the major overhaul and the relief valve work were not cross-referenced one to the other. However, these were not isolated incidents of failures in the PTW system. The operator's written PTW procedure did not mention the need to cross-reference permits where one piece of work may affect another. It made no reference to methods of isolation or locking-off valves to prevent inadvertent recommissioning. It was essential that operating staff work exactly to the written procedure but that did not take place on Piper Alpha. Numerous errors were regularly made in the way permits were completed. Multiple jobs were undertaken on a single permit. It was common practice for maintenance supervisors to leave permits on the control room desk without discussion as to the state of the work with process staff. Process supervisors frequently signed off permits before having the state of the equipment checked out.

It is essential that staff who are required to operate a PTW system are thoroughly trained in all its aspects. That was not the case on Piper. Neither the operator's own staff or contractors' supervisors were provided with formal and regular training to ensure they operated the system as laid down. All training was 'on the job', that is learning from other supervisors. That has a part to play but as the sole method, it suffers from the crucial weakness of perpetuating or accumulating errors.

The leak of condensate that led to the explosion that initiated the disaster would not have occured if there had been a better PTW system; the deficiencies were due, in part, to a lack of training in the operation of that critical safety system.

17.2 The oil fire

The overpressure from the explosion in the condensate area blew down the fire wall separating it from the section of plant containing equipment to extract crude oil. Containment was breached and there was an immediate and large oil fire. It was that fire and the resultant engulfing of the platform in thick smoke which so hampered attempts to escape.

Evidently, in the design stage of the platform, it had been judged that a fire could occur due to a loss of containment in the condensate section. The probability of that happening was sufficiently likely that fire walls were installed between the various sections of the unit. It is difficult to imagine that, in the light of that assessment, an explosion in the condensate area was any less likely. Nevertheless, Piper Alpha had no explosion walls either side of the condensate area. However, the more crucial deficiency was that, at the design stage, there had not been a systematic and, where necessary, quantified assessment of the major hazards which could occur on the platform. No hazard and operability review had been made of the design.

The oil fire was larger and persisted for longer than the inventory of crude oil on Piper would sustain. This was due to oil leaking back into the fire from the common oil export line, either because the automatic isolation valve had been damaged by the explosion or the oil line had been fractured. Pressure was maintained in the common export oil line because the two platforms which shared it did not decide to begin shutting down production until about one hour after the explosion on Piper. Those delayed decisions were in part due to the lack of any communication from Piper itself, all its radio facilities having been destroyed by the explosion. Additionally there had been no practising of the type of response which should be undertaken in the event of an·emergency involving fire or explosion on one of the three linked platforms, another example of deficiency in providing adequate safety training.

17.3 The gas fire

The gas pipelines connecting Piper Alpha to the neighbouring platforms came into Piper itself in the area of the major oil fire. The pipelines operated at pressures up to 120 bar (1750 psi) were 20 to 56 km long and between 400–460 mm diameter. As such they contained a massive inventory of gas which could not be blown down quickly. The pipeline terminals on

Piper were weakened by the oil fire and successively ruptured, engulfing the platform in a ball of fire and making escape for those on board if not impossible extremely hazardous.

The hazard presented by the gas inventory was well understood by the operating management. It had been highlighted just 12 months earlier in a report which considered the possible types of fires which could occur on Piper and their consequences. Nevertheless, no specific provision was made to protect the pipeline end terminals where they joined the platform against being weakened and failing and allowing the pipeline contents to escalate an on-board fire. Either fireproofing or dedicated water sprays could have been used, although each presented operational problems. Either might have delayed the pipeline failures on the night of the disaster and thus provided more time for the crew to effect their escape.

In this example the deficiency was one of satisfactory assessment of the probability of a known major hazard and its consequences and taking whatever precautions were available and justified, if not to avoid it, to at least mitigate its impact.

17.4 Fire-fighting

It is inescapable that any offshore platform has to tackle an emergency with the facilities installed in the platform itself and with the onboard crew. In contrast to an onshore process plant, local emergency services cannot be called in to help. A well-trained, powerfully equipped and ably led fire service cannot be on site in just a few minutes. Some offshore platforms, and Piper was one, do have stand-by fire-fighting ships that can pump water onto the unit but this has little effect.

The primary fire-fighting equipment on Piper was a water deluge system which sprayed a defined quantity of water over all areas which had hydrocarbon-containing equipment. It was fed from electrically driven water pumps which unfortunately were put out of action when the initial explosion destroyed the main power supply. The risk of power failure had been foreseen in the platform design and back-up diesel driven pumps were provided to start up automatically if the main electrical pumps did not work. However, on the night of the disaster those diesel pumps were on manual start, and could not be reached due to the oil fire, although two operators bravely tried. They were never seen again. The pumps were on manual start not in error but by deliberate decision. The pumps used sea water and if they started up while divers were near their suction intakes the divers could be at risk. This was certainly a hazard but the practice on Piper was to have the stand-by diesel pumps on manual start whenever the divers were in the water as opposed to when diving was taking place close to the suction intakes. In the summer months diving occurred almost every night and thus for half of each day the operability of the fire-fighting system was inhibited, and inhibited in an unnecessary and dangerous way. The practice on other platforms was to have the stand-by pumps on manual start only when diving was taking place near the pump intakes, which was infrequent.

Even if the diesel water pumps had started it is unlikely that the water deluge system would have been fully effective. Part of the system would have been destroyed by the explosion but the whole system had suffered from the blocking of the deluge heads, caused by sea water corroding the pipework and the corrosion products lodging in the spray heads. Various methods had been tried to overcome the problem but without success and finally a decision had been taken to replace the deluge pipework in non-corrosive material. At the time of the disaster only part of the system had been replaced. What was unacceptable was that the problem of blocking spray heads had been identified four years before the disaster and so for that whole time a system critical to the platform safety was less than fully effective.

The deficiencies in the fire water deluge system were not ones of lack of recognition of a problem but of quality of decision on the diesel pumps and timely resolution on the blocking spray heads.

17.5 The accommodation block

It was understandable that the crew should assemble in the accommodation block in the face of the emergency. Many were already there at the time of the explosion. Others on duty made their way there prior to the first gas pipe failure as they expected to be rescued by helicopter, the normal mode of transport to and from the platform. The helicopter pad was located on the roof of the accommodation. At first conditions were not too bad. There was still battery driven emergency lighting and the smoke in the atmosphere was light. However, after the emergency lighting failed panic set in. Smoke became much thicker and eventually intolerable. Nearly all those who died in the accommodation did so as a result of inhaling smoke and gas.

The accommodation block was designed to resist fire for some time but not specifically to prevent smoke ingress. However, the ventilation system dampers were installed to shut on high temperature and so that was not the means by which the smoke entered the accommodation. Smoke ingress was due to people opening doors to assess the possibility of escape and, crucially, to some fire doors being normally hooked open to facilitate movement. In any emergency it should have been axiomatic to shut fire doors and exercise discipline in the opening of others.

It should have been apparent within a few minutes of the oil fire developing, and certainly once the gas fire enveloped the platform, that there was no possibility of a helicopter being able to land. The only chance of survival, however hazardous, for those in the accommodation was to fight their way through the smoke and flames, jump into the sea and hope to be picked up by one of the ships which had come to help. Unfortunately, no orders to do that were given by any of the senior personnel. No-one can forecast how anybody will react under such stressful circumstances but it was evident that neither the installation manager nor any of his senior supervisors had received thorough training in how to lead in a major emergency or had undergone regular, simulated emergency exercises. Some men of their own initiative did leave the block and were rescued but the majority stayed inside where death was inevitable

17.6   Safety auditing

The way that any management ensures that its decisions on safety procedures are carried out in practice is to regularly audit operations. The deficiencies outlined in the previous sections became readily apparent in the inquiry into the accident so why were they not apparent to the operating management? Certainly, the time devoted to safety auditing should have been sufficient.

Compliance with the PTW procedure was monitored each day and had been one subject of a parent company audit just 6 months before the disaster. No deficiencies were reported. An annual fire safety audit was undertaken but its report had not referred to the problem of blocking deluge heads. An audit a few years earlier had picked up the practice of putting the diesel driven fire pump on manual start and recommended that should only be done when divers were working near the pump intakes. That recommendation had never been implemented, nor had the audit team followed up to check that it had been.

Clearly, there was no shortage of auditing of the Piper platform and the way it was being operated. What was deficient was the quality of that auditing. Not only were departures from laid-down procedures not picked up, but the absence of critical comment in audit reports lulled the senior management into believing that all was well.

17.7   Lessons from the disaster

There are many lessons to be learnt from the events and background to the tragedy of Piper Alpha. Permit-to-work systems must involve a secure method of locking off valves to prevent inadvertent opening; there must be a systematic assessment of all potential major hazards at the design stage; interactive effects in emergencies between linked operating units must be thought through; a system for the timely resolution of faults in safety critical equipment must be part of normal operating management. However, history shows that the detailed circumstances of any major accident do not repeat themselves. Each is unique in its train of events and it is for that reason that it is essential to examine the root causes of an accident. From that approach there are four key lessons to be drawn from the Piper disaster which apply to any large-scale process operation.

(1) *Management is responsible.* All the deficiencies outlined in the previous sections had one thing in common; they were the responsibility of management and management alone. Only an operator's management has the knowledge, the power to act and the legal responsibility to ensure a safe environment in which employees work. Management is a wide term but in relation to safety it is the line management which is responsible. Specifically, safety is not the responsibility of the Safety Department. That department has important contributions to make, such as supporting and advising line management, monitoring developments in the industry or in regulatory approach and may be used to carry out safety auditing but it cannot be responsible for safety performance. That has to be solely the line management from design through the operating life of any plant.

(2) *A systematic approach is required.* The deficiencies on Piper Alpha were failures in systems. Either there was a system but it was inadequately designed and executed, for example, the PTW system, or there was no system where one should have existed, for example, the lack of a systematic method for assessing major hazards or the lack of a system for training in inter-platform emergencies. Good safety performance cannot be achieved by a 'hit or miss' approach based on the experience and imagination of staff. There has to be a carefully thought out and rigorously executed suite of safety management systems.

(3) *Quality of safety management is critical.* Not only does any operation have to have the right safety systems but they have to be quality systems, that is they have to meet the defined requirement each and every time and each and every day. On Piper the installation manager was right to consider the danger to divers posed by the automatic start of the diesel fire pumps but his decision was lacking in quality. The PTW system was not a quality system. Some or perhaps most of the maintenance tasks were carried out safely but the deficiencies in the design and particularly the operation of the system prevented it ensuring that all maintenance work would be achieved without danger to staff or the installation.

(4) *Auditing is vital.* In any organisation it is an eternal concern for management to make sure that its decisions and procedures are carried out exactly as it has determined. Nowhere is that more important than in the area of safety. For that reason it is essential that a regular and thorough auditing system is one of the operation's safety management systems. On Piper there appeared to be sufficient effort put into safety auditing but it was evident that it was not of the right quality as otherwise it would have picked up beforehand many of the deficiencies which emerged in the inquiry into the disaster.

Event	Recommendations for prevention/mitigation
New regulations and public concern compelled off-shore operators to improve standards and demonstrate that they are safe	
▲	
167 men killed, many by inhalation of gas	
▲	
Fire worsened; many men did not know what they should do	
▲ ◄—————————————	**Better training especially of supervisors**.
Fire-fighting ineffective as pumps fail to start and spray heads are blocked	
▲ ◄—————————————	Put pumps on manual control only when essential *Replace spray heads in non-corrosive material (under consideration for 4 years)*.
Gas lines ruptured	
▲ ◄—————————————	Fire-protect gas lines.
Fire fed by fresh oil export line	
▲ ◄—————————————	**Provide better communication with other platforms. Plan for emergencies**.
Compartment walls blown down; fire spreads	
▲ ◄—————————————	Provide explosion-resistant walls. Carry out hazops and QRA on new designs.
Explosion and fire	
▲	
Condensate leaks from loose flange	
▲	
New shift started up pump not knowing relief valve was removed	
▲ ◄—————————————	**Provide better hand-over between shifts. Enforce PTW system**.
PTW for overhaul signed off but process team not told that relief valve was removed	
▲	
Relief valve removed; open end loosely blanked	

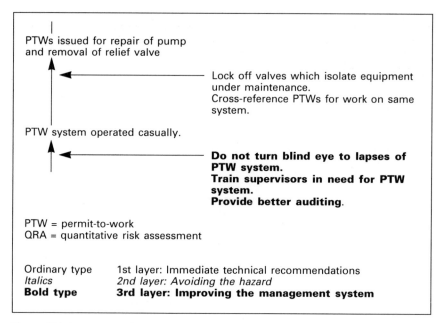

PTW = permit-to-work
QRA = quantitative risk assessment

Ordinary type	1st layer: Immediate technical recommendations
Italics	*2nd layer: Avoiding the hazard*
Bold type	**3rd layer: Improving the management system**

Figure 17.3 Summary of Chapter 17 – Piper Alpha

17.8   Conclusion

As we saw in Section 8.8, the explosion at Flixborough in 1974 resulted in the UK in the CIMAH (Control of Industrial Major Accident Hazard) Regulations, which require companies that store or process large quantities of hazardous materials to demonstrate that they can do so safely. In the same way the Piper Alpha explosion resulted in new regulations for the offshore oil and gas industry which require operators to do the same. In the offshore industry companies are required to carry out a numerical assessment of the risk; onshore this is encouraged but is not normally required.

While the lessons above are the main ones to emerge from considering this tragic accident it is clear that no single act or omission was responsible for the deaths of so many men. As so frequently happens a major accident is the result of a sequence of events which, each in themselves, is unlikely. Equally no single individual or small group can be said to be responsible for the disaster. Perhaps the most important lesson of all that can be drawn is that the sum and quality of our individual contributions to the management of safety determines whether the colleagues we work with live or die.

References

1   Cullen, W., D., *The Public Inquiry into the Piper Alpha Disaster*, Her Majesty's Stationery Office, London, 1990.

Chapter 18

The King's Cross underground railway station fire

A little fire is quickly trodden out, Which being suffered, rivers cannot quench.

Shakespeare, *Henry VI*, Act 4, Scene 6

Thirty-one people were killed and many more injured by a fire at King's Cross underground railway station in London on 18 November 1987. The report of the official inquiry[1], on which this chapter is based, was very critical of the management of London Underground, as can be seen from the amount of bold type in the summary (Figure 18.2) and the ways in which the various events could have been prevented or mitigated. Quotations from the report are in *italics*.

18.1   The immediate causes

The immediate cause was a lighted match dropped by a passenger on an escalator. Although smoking was not allowed on the underground, including the stations, passengers often lit up as they ascended the escalators on the way out of the stations and this was tolerated.

The match fell through a gap between the treads and the skirting board. These gaps were caused by sideways movement of the treads and although cleats were supposed to prevent (or reduce the likelihood of) anything falling through the gap over 30% of the cleats were missing.

The match set fire to an accumulation of grease and dust on the running tracks. The tracks were not regularly cleaned and had not been designed so as to minimise the accumulation of dust or to make cleaning easy. Non-flammable grease was not used. The fire soon spread to the wooden treads, skirting boards and balustrades. More modern escalators were made from metal but replacement of the old ones was slow.

The fire was first noticed by a passenger at 19.29. At 19.30 another passenger saw smoke two-thirds of the way up and a glow underneath. He pressed the emergency stop button at the top and shouted to people to

Figure 18.1 The top of the escalator at King's Cross after the fire. (Reproduced by permission of the Health and Safety Executive.)

get off the escalator. The Fire Brigade was summoned at 17.34 by a British Transport policeman. By this time the flames were 3–4 inches (75–100 mm) high. The first fire engine arrived at 19.42. At this time the fire was described as being of the size that might have been created by a large cardboard box. Suddenly, at 19.44 or 19.45, before the firemen had time to apply any water, the fire suddenly and rapidly spread into the ticket hall at the top of the escalators accompanied or preceded by thick black smoke. All but one of the deaths occurred during the period immediately following. In one or two minutes the fire spread from what seemed to be an innocuous one into an inferno. It belched into the ticket hall like a flame gun[2]. Figure 18.1 shows the top of the escalator after the fire.

The practice in London Underground was to call the Fire Brigade only when a fire seemed to be getting out of hand. If smoke detectors had been installed, and the Fire Brigade called as soon as they sounded, then the extra five minutes or so might have given the firemen enough time to apply water. In addition, water could be applied automatically.

The escalator running tracks were fitted with a water spray system which had to be activated manually. During the eight minutes which elapsed between calling the Fire Brigade and their arrival the London Underground inspector on duty walked past the unlabelled water valves. He was not a regular inspector, only an acting one, and did not know the location of the valves. In addition, to quote the official report, . . .*his lack of training and unfamiliarity with water fog equipment meant that his preoc- cupation with the fire and smoke led him to forget about the system or the merits of its use* (p. 62). London Underground employees were promoted largely on the basis of seniority and had little or no training in the action to take in an emergency.

The other staff on duty showed a similar lack of training and ability. Their reactions were haphazard and uncoordinated. They were *woefully inequipped to meet the emergency that arose* (p. 67). Thus between the time the fire was detected and the time it erupted into the ticket hall, a period of about 15 minutes, people were allowed to go through the ticket hall and up and down other escalators (which were located in separate tunnels). The station could have been closed and passengers on the platforms evacuated by train. (Unfortunately an alternative exit to another line was closed in the evenings.) As we shall see this lack of action was due to the view, shared by all concerned, that fires on escalators were not a serious problem.

18.2 The underlying causes

18.2.1 Fires are inevitable and trivial

A major underlying cause was the belief, held at all levels, that escalator fires were inevitable and would not cause serious injury to passengers. This view was positively encouraged by the management who insisted that they should be called smoulderings to make them seem less serious. With this background it is understandable that the acting inspector, when informed of the fire, did not inform the station master or the line controller (p. 63).

Fires were certainly numerous: 400 between 1958 and 1987, an average of twenty per year. Some caused damage or delays and stations had to be evacuated but no-one was killed though some passengers were taken to hospital suffering from smoke inhalation (p. 45).

Safety advisers are often told, 'It must be safe as we have been doing it this way for twenty years without an accident'. However, the fact that no one has been killed in twenty years is relevant only if a fatality in the 21st year is acceptable. In fact, twenty years without a fatality does not prove that the probability is less than one in twenty years. All it proves is that we can be 86 percent confident that the probability is less than once in ten years[3].

I am often amazed at the statistical innumeracy of some engineers. Unfortunately, if it is convenient to believe something, scientific training and commonsense rarely stand in the way.

In the incident described in Chapter 4 the attitude was: don't worry about leaks; they won't ignite. Similarly, at King's Cross the attitude towards escalator fires was; don't worry: they won't escalate.

18.2.2   Unsuitable personnel and poor training

On the night of the fire nine railway workers, a relief inspector, a station manager and a relief station manager were on duty (excluding booking clerks and the staff on the adjoining Metropolitan line station, supervised by the same station manager). Four of the railway workers were restricted on medical grounds: one was brain-damaged, one had a heart condition, one had had a nervous breakdown and one had a respiratory problem. One worker who should have been present had left early without permission and another had leave to attend hospital but had not returned. Three were taking extended meal breaks.

Training on fire-fighting amounted at the most to one day but some men had not even had that. There was no re-training or on-the-job training. London Transport set great store by the use of trains to evacuate passengers in an emergency but this method was not used as there were too few staff available to cover the six platforms and those that were available were not trained. Decisions about evacuation were made by the police who decided, at 19.40, to evacuate people from the platforms via the other escalators and the ticket hall. The public address system was never used.

Indeed the operations director accepted that it was likely that there was nobody who had a nationally recognised qualification at King's Cross station on 18 November 1987, when they were responsible for perhaps £40 million worth of assets and a quarter of a million passengers (p. 30).

18.2.3   Failures to learn from the past

The recommendations made following earlier escalator fires, and a serious platform fire in 1984, had not been carried out. These recommendations included:

- Calling the Fire Brigade as soon as a fire was reported.
- More frequent and thorough cleaning of escalator running tracks.
- Replacement of wood by metal.
- Moving the valves which operated the water fog equipment into areas less likely to be affected by smoke.

Why were these recommendations not carried out? No one took a considered decision not to do so. Instead no-one 'owned' the recommendations and accepted responsibility for seeing that they were carried out. *There was no system in place to ensure that the findings and recommendations from such inquiries were properly considered at the appropriate level. . . . there was not sufficient interest at the highest level in the inquiries. There was no incentive for those conducting them to pursue their findings or recommendations.*

London Underground's failure to carry through the proposals resulting from earlier fires. . . was a failure which I believe contributed to the disaster at King's Cross (p. 117).

As far back as 1901 the Chief Inspecting Officer of Railways had recommended, in his report on a fire in Liverpool, that as little wood as possible should be used on underground stations on electrified lines[4].

18.2.4   The management of safety

The safety professionals in London Underground were junior in status and concerned only with employee safety, not passenger safety.

The influence of an adviser depends primarily on the quality of his advice but it also depends on his status within the organisation. If the senior safety adviser is responsible to a junior or middle manager he will find it difficult to influence decisions. *For example, the chief fire inspector, Mr Nursoo, found the same problems of poor housekeeping and electrical wiring in escalator machine rooms year after year. He duly reported this to his superiors but told the Court that he was powerless to require action to be taken (p. 127).*

Safety professionals were not concerned with passenger safety as London Underground believed that *passenger safety was inextricably intertwined with safe operating practices* (p.116). The same argument is often heard in other industries. Of course, managers, not safety advisers are responsible for safety, but safety advisers are needed to advise, monitor, provide information, train and assist with specialist studies and surveys. They should not wait until they are asked for advice but should say what they think should be done[5].

In the same way managers are responsible for profits and people but they still need accountants and personnel officers to advise them.

At senior levels, as shown in Section 18.2.3 on learning the lessons of the past, there was no clear responsibility for safety, no system for ensuring that what should be done was done and safety did not receive, at least so far as stations were concerned, the attention it deserved.

Like the organisation described in Chapter 4, *Compartmental organisation resulted in little exchange of information or ideas between Departments, and still less cross-fertilisation with other industries and outside organisations. . . .it undoubtedly led to a dangerous, blinkered self-sufficiency which included a general unwillingness to take advice or accept criticism from outside bodies (p. 31).*

18.3   Was there no watchdog?

Many readers may be surprised that a publicly owned organisation was allowed to get into the state described. Was there no watchdog?

The Health and Safety at Work (HSW) Act 1974 requires employers to provide a safe plant and system of work and adequate instruction, training and supervision, so far as is reasonably practicable. The Health and Safety Executive (HSE), which is responsible for monitoring and enforcing the Act, had appointed the Railway Inspectorate, a body founded in 1840, as its agents. The Railway Inspectorate was criticised in the report on the fire for believing that the 1974 Act placed no new duties on the railways and that ... *a proper observance by the railways of the statutory duties placed on them by railway legislation and Transport Acts would equate to a discharge of their duties under Section 3* (p. 145) (that is, their duties to the public). This, however, is not the spirit of the HSW Act. The basis of the Act is that a plant cannot be made safe just by following a set of rules. (For an example see Section 20.1.) Employers must be continually on the lookout for new hazards and new ways of controlling existing hazards.

The Railway Inspectorate is now part of the Health and Safety Executive.

18.4   Did the report go too far?

The official report included 157 recommendations which were accepted by London Underground and the other organisations involved such as the London Fire Brigade. They were classified as very important, important, necessary and suggested.

The chairman and managing director of London Underground (and the chairman of the organisation to which it was responsible, London Regional Transport) resigned when the report was issued. In a speech in 1992 the new managing director accepted the 'damning criticism of the way we were managing the company at that time'. However, he felt that in some places the report had gone too far as it had failed to use quantitative risk assessment (QRA) or cost–benefit analysis and had made recommendations that would produce little benefit. For example, London Underground faced expenditure of £100 million pounds over a year to comply with fire precaution regulations to save about a fifth of a life per year; 'We don't think that's good value for money.' After the fire London Underground had 'brought London virtually to its knees by attacking every escalator and tearing out all the wood'. Intuitively, that had seemed a good idea but calculations showed that while this would reduce the probability of a serious escalator fire from once in six years to once in nine, installation of sophisticated sensors and automatic sprinklers would reduce the probability to once in a thousand years.

The new managing director praised QRA for compelling people to face the setting of safety spending priorities and the valuation of human life and accused media persons, politicians and others of publicly implying infinite value for each life. Yet motoring, flying – indeed all activity – would cease if we did not accept a trade-off between risk and benefit. Nevertheless, QRA did not supersede judgement but should lie alongside it[6].

Similar criticisms were made in a report by Brian Appleton, one of the assessors to the Piper Alpha inquiry and author of Chapter 17 of this book[7]. The report was produced at the request of the HSE following an incident on London Underground in 1992 when two suspect briefcases were found in a train. Seven trains were stopped in tunnels during a morning rush hour as there were more trains on the line than there were stations to stop them at. It took five hours to evacuate all 6000 passengers seventy of whom were taken to hospital with heat exhaustion. Smoke from a short circuit on one of the trains added to the confusion and if it had developed into a fire the result might have disastrous. The briefcases turned out to be harmless pieces of lost luggage.

The report says that closure and evacuation of stations may not always be the right response. It recommends that railway staff are given training, similar to that given to airport staff, to help them assess the seriousness of bomb warnings. On fire prevention the report is more positive. It says that as a result of the action taken since 1987 the situation has been transformed and fire prevention should no longer claim a lion's share of resources. Instead QRA should be used to assess priorities. The existing legislation, based on regulations which must be followed, should be replaced by one based on the quantitative assessment of risk.

18.5   Conclusions

Many readers may be shocked by the conditions disclosed in the official report. Let us keep matters in perspective. Passengers on London Underground are about as safe as they are at home and fifty times safer than they are in cars[6]. Thirty-one people are killed every few days on the roads without any public inquiry, press report or demands for action.

The overall impression given by the King's Cross report is that London Underground ran trains very efficiently but were less interested in peripheral matters such a stations. In the same way many process plants give services such as steam, water, nitrogen and compressed air less than their fair share of attention and they are involved in a disproportionate number of accidents. The same is often true of outlying or overseas plants. Who would have thought, beforehand, that Bhopal (Chapter 10) would have occurred in a company such as Union Carbide or Flixborough (Chapter 8) in a company such as Dutch State Mines?

Rot starts at the edges. To avoid a King's Cross in your organisation give your service, support and outlying activities the same degree of attention as the heartland and the flagship.

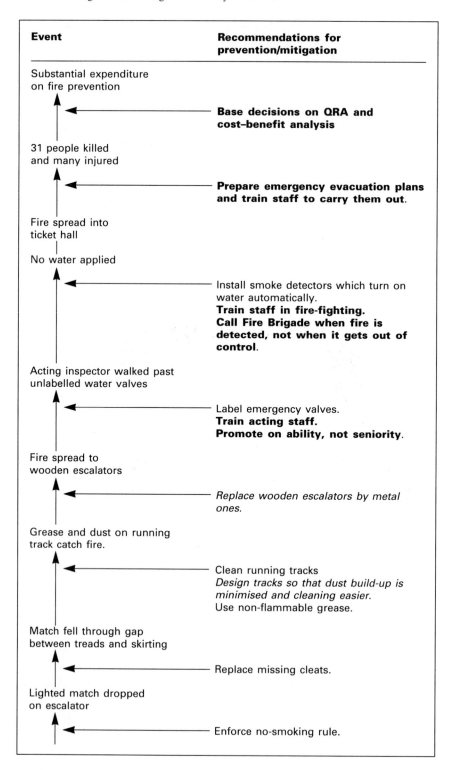

Event	Recommendations for prevention/mitigation
Substantial expenditure on fire prevention	Base decisions on QRA and cost–benefit analysis
31 people killed and many injured	Prepare emergency evacuation plans and train staff to carry them out.
Fire spread into ticket hall	
No water applied	Install smoke detectors which turn on water automatically. Train staff in fire-fighting. Call Fire Brigade when fire is detected, not when it gets out of control.
Acting inspector walked past unlabelled water valves	Label emergency valves. Train acting staff. Promote on ability, not seniority.
Fire spread to wooden escalators	Replace wooden escalators by metal ones.
Grease and dust on running track catch fire.	Clean running tracks Design tracks so that dust build-up is minimised and cleaning easier. Use non-flammable grease.
Match fell through gap between treads and skirting	Replace missing cleats.
Lighted match dropped on escalator	Enforce no-smoking rule.

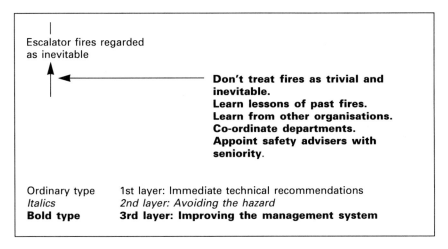

Escalator fires regarded
as inevitable

**Don't treat fires as trivial and
inevitable.
Learn lessons of past fires.
Learn from other organisations.
Co-ordinate departments.
Appoint safety advisers with
seniority**.

Ordinary type   1st layer: Immediate technical recommendations
Italics   *2nd layer: Avoiding the hazard*
Bold type   **3rd layer: Improving the management system**

Figure 18.2 Summary of Chapter 18 – King's Cross underground railway station fire

References

1 Fennell, D., *Investigation into the King's Cross Underground Fire*, Her Majesty's Stationery Office, London, 1988.
2 Crossland, B., *Lessons from the King's Cross Fire*, Synopsis of a lecture, Fellowship of Engineering, London, 1989.
3 Kletz, T. A.,*Improving Chemical Industry Practices – A New Look at Old Myths of the Chemical Industry*, Hemisphere, New York, 1990, Section 33.
4 Hall, S., *Railway Detectives*, Ian Allan, London, 1990, p. 65.
5 Kletz, T. A., *Lessons from Disaster – How Organisations have No Memory and Accidents Recur*, Institution of Chemical Engineers, Rugby, UK, 1993, Section 7.2.
6 Conway, A., *Atom*, No. 420, Feb. 1992, p. 9.
7 Appleton B., *The Appleton Report*, Her Majesty's Stationery Office, London, 1992.

Chapter 19

Clapham Junction – Every sort of human error

We have lived with the disgrace and now have to show that this cannot happen in the future.

Sir Bob Reid, Chairman of British Rail[1]

Just after 8.00 am on 12 December 1988 a crowded commuter train (B) ran head on into the rear of another train (A) which was stationary in a cutting near Clapham Junction railway station in South London. The moving train (B) veered to the right and struck a train (C) coming in the opposite direction. Thirty-five people were killed and nearly 500 injured, sixty-nine seriously. All those killed were travelling in the front two coaches of train B (see Figure 19.1).

The stationary train (A) had stopped because a signal suddenly changed from green to red when the train was almost on top of it. In accordance with instructions the driver stopped at the next signal to report the fault by telephone to the signalman. He assumed, as he had every right to do, that the track-circuiting would detect the presence of his train on the tracks and would keep the previous signal at red. Unfortunately there was a fault in this signal, a wrong side failure; it was green when it should have been red; it was inviting the next train (B) to pass it and enter the section already occupied by the stationary train (A) (Figure 19.2).

The official report[2], on which this chapter is based (quotations are in *italics*) and which is summarised in Figure 19.3, established the immediate and underlying causes of the accident, all of them due to deficiencies in the management. Between them they illustrate many different sorts of human error[3].

19.1 The immediate causes – a 'mistake' and a slip

During the two weekends before the accident a new signalling system had been installed and one of the signalling technicians failed to follow the correct procedures. After disconnecting a wire which was no longer

Figure 19.1 The second coach (buffet car with loose seats) of the second (moving) train involved in the railway accident at Clapham Junction. (Reproduced by permission of London Fire Brigade.)

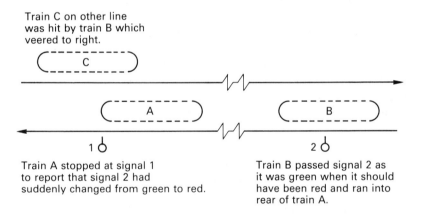

Train C on other line
was hit by train B which
veered to right.

Train A stopped at signal 1
to report that signal 2 had
suddenly changed from green to red.

Train B passed signal 2 as
it was green when it should
have been red and ran into
rear of train A.

Figure 19.2 The Clapham Junction railway accident

required he failed to cut it back so that it could no longer reach its old position; he did not tie it out of the way; and he used old insulating tape instead of new to insulate the bare end.

These failures to follow the correct procedures were not isolated incidents. The technician was following the working practices he had followed for sixteen years. On this occasion his immediate supervisor did not pick up the wrong methods of working; more seriously, none of his supervisors had ever picked them up. The errors were not 'violations',

Event	Recommendations for prevention/mitigation

Massive rescue operation

← Many suggestions for improving the response of the emergency services.

Train ran into stationary
one in front; 35 killed and
nearly 500 injured

← *Strengthen railway stock.*

Driver stops to report signal fault

← Provide radio contact.

'Wrong side' signal failure

No checks by supervisor
or testing team

← **Appoint elite group of testers. Check that decisions are carried out.**
Do not let supervisors do so much work themselves that they cannot check others.
Avoid continuous multiple checking as each checker may rely on the others.

Slips by signalling technician
(wire disconnected at one end only
and other end not insulated)

← **Do not let people work for long periods without a day off.**
Always check critical jobs.

'Mistakes' by signalling technician
(disused wires not cut back
or secured; old tape re-used)
over many years

← **Check everyone's work from time to time.**
Audit training and workmanship.
Always check critical jobs.

Instructions not received or read

← **Explain and discuss new instructions.**
Write instructions which help the reader rather than protect the writer.

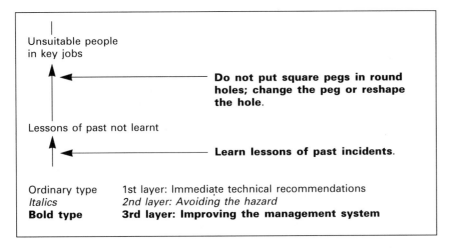

Figure 19.3 Summary of Chapter 19 – Clapham Junction

deliberate decisions not to follow instructions, but genuine mistakes, the result of poor training and poor supervision; the technician thought he was doing the job correctly.

It might be argued that if the technician had read the instructions he would have discovered that his method of working was wrong. This is unrealistic. Few of us, carrying out routine tasks without apparent trouble, ever stop to read the instructions to see if we are doing them correctly. In the present case it is doubtful if the technician was ever given written instructions; he was trained on the job. At the official inquiry he was ready to accept almost total responsibility for the accident. The report says, *In that he happens to be wrong and to do himself an injustice. [He] had the right to expect that those who monitored, those who supervised, those who managed staff at his level would have had sufficient regard for safety to have ensured*:

(i) *that the quality of work carried out on the installation of new works was of an acceptable standard; and*
(ii) *that staff were given appropriate instruction and training to ensure improvement of standards and the maintenance of those improved standards (§8.6,7).*

In addition to his habitual errors the technician had a lapse of attention which resulted in two further errors *which had nothing to do with his normal practice and were totally uncharacteristic of his work* (§8.17): he disconnected a wire at one end only and did not insulate the other bare end.

We all make slips and have lapses of attention from time to time. They are inherent in human nature and cannot be avoided. It is useless to tell people, as we often do, to take more care or to keep their mind on the job. As the report states, *Any worker will make mistakes during his working life. No matter how conscientious he is in preparing and carrying out his work, there will come a time when he will make a slip. It is these unusual and infrequent events that have to be guarded against by a system of independent checking of his work* (§8.26). There was no such system.

The technician may have made the slips because he had been working seven days per week for many weeks and this produced a *blunting of the sharp edge of attention which working every day of the week, without the refreshing factor of days off, produces* (§8.26). Alternatively, his concentration may have been broken by an interruption of some sort. He was not physically tired.

The quotations illustrate the sea-change in our attitude towards human error that has occurred in recent decades. It is no longer acceptable to blame the man at the bottom, who cannot pass on the blame to someone below him, for errors that could be prevented by better training and supervision or for the slips and lapses of attention that inevitably occur from time to time.

19.2   Why was there no checking?

There were supposed to be three levels of checking, wire counts by the technician, the supervisor and a tester.

The supervisor was so busy leading an outside gang and carrying out manual work himself that he left no time for checking the technician. He *over-involved himself in carrying out tasks rightly the responsibility of technicians and under-involved himself in the tasks essential to his role as a supervisor* (§8.36). It is, of course, a common failing at all levels, for people after promotion to try to continue to carry out their old job as well as their new one, to the detriment of both. In this case the supervisor's excuse was that it was the usual practice for technicians to carry out wire counts *as they went along*, and that should be sufficient. Neither the technician nor the supervisor recognised the importance of a final wire count at the *end of the job*. Neither of them had seen instructions issued in 1985 and 1987.

Sending instructions to people through the post is of limited value. Even if they receive them they tend to put them aside to read when they have time. New instructions should be explained to people and discussed with them and checks should be made to confirm that they are being followed. If this is not done people may carry on as before.

The tester (acting Testing and Commissioning Engineer to give him his full title) did not carry out a wire count, only a functional test. He thought that was all that was required. He had been doing the job for only a few months and had received no real induction training. *He was doing a job he had no wish to do at a place where he had no wish to be. If he had little liking for the job, he had less enthusiasm* (§9.5).

The supervisor's superior was also unhappy in his job. After 42 years service and with only another 18 months to go *he was pitched in the twilight of his career into work that was foreign to him* (§16.18). He turned a blind eye to irregularities, not wishing to interfere with the accepted way of doing things.

The senior managers were not unaware that there were faults in the testing system. Several years earlier the Signal and Telecommunications Engineer for the Southern region of British Rail had identified the problem (poor testing) and arrived at a solution (new instructions, a new cadre of testers, and a new Regional Testing Engineer charged with the job of raising standards). *The difficulty was that having identified the problem and arrived at solutions, he then turned his attention to other things and made the dangerous assumption that the testing would work and that the problem would go away. In fact it did not. No cadre of testers ever came into being... No training course was ever run.* The new engineer *did not raise the raise the overall standard of testing throughout the region. The urgently-needed full new testing instruction did not even appear for a year-and-a-half and was then virtually ignored...* (§16.65).

The answer to the question in the title of this section is therefore that there was no testing because:

- many of the staff were unsuited to the jobs they were given;
- the training and supervision were poor;
- there was a failure to follow up.

In short, *There was incompetence, ineptitude, inefficiency and failure of management...* Responsibility was shared by the three individuals at the workface who made specific errors and *those who allowed an originally sensible and workable system to degenerate into an incompetent, inept and potentially dangerous way of working* (§8.45–§8.46).

A point not made in the official report is that three checkers may be less effective than one. Each checker may assume that the other two will carry out their checks and so it does not matter if he skips them for once. 'For once' can soon become habitual. In contrast, a single checker, encouraged by occasional spot checks by others, knows that everyone relies on him. In the present case, however, checking did not occur because the three men concerned did not know that it was required.

19.3 Communication failures

A provisional instruction on testing, requiring three wire counts, was issued in 1985 and reissued, in a final form, in 1987. The technician and his supervisor said that they had never seen it; other supervisors said the same. The tester saw the 1985 instruction but thought that it was merely a discussion paper; he never saw the 1987 version (§8.34 and 9.7). The tester's superior received the 1987 instruction but *had no time to sit and study it. . . He said he merely filed it and carried on with the mounting volume of work he had* (§9.23).

The Regional Testing Engineer also did not realise that the new instruction was in force. *It is a matter which the Court can only look upon with both alarm and horror. . .that the man in overall charge of the testing of new works for the whole of Southern Region should have arrived at a conclusion that a Departmental Instruction which had as its very title 'TESTING OF NEW AND ALTERED SIGNALLING' and had been formally and properly issued, was not in force, and that he could have persisted in that view for a period of one and a half years between the issue of that document and the accident* (§9.44).

As already stated, Clapham Junction shows very clearly the futility of sending instructions through the post. Unless they are explained and discussed we cannot know that they have been received, understood and accepted.

Were the communication and other failures described in this chapter typical of British Rail at the time? In 1991/2 the Health and Safety Executive investigated falls from railway carriages. Their report[4] shows that there were written procedures for the maintenance of door locks but *those undertaking the work at shop floor level were not aware of the existence of the detailed instructions or did not have copies* (§106) and *as a last resort, fitters used mallets on the lock handle to effect a final alignment of the lock and handle. This can cause twisting of the handle spindle and could affect future alignment* (§109); *. . .it was apparent that British Rail had not addressed these major maintenance problems in a co-ordinated manner and many of the local variations were undertaken in ignorance of practice elsewhere* (§110).

However, poor maintenance was not the main reason why the locks failed to function. The design was *overly sensitive to small variations in dimensions and conditions, making fitting and adjustment for fault free operation very difficult* (§236).

19.4 Failures to learn from the past

Five wiring failures, similar to those made at Clapham Junction, occurred elsewhere in the Southern Region in 1985, fortunately without causing a serious accident. *They should have provided the clearest possible lessons that all was not well in the ST Department. They should have been used as clear pointers to what was going wrong in the installation and testing practices on Southern Region* (§9.52). Instead all they produced was a new provisional instruction, not made permanent for 18 months, *which was never to be properly understood or implemented and was not to change any of the existing practices for the better, . . .not a single training course on testing in the three years before the Clapham Junction accident [and] the appointment of the Regional Testing Team. .with the objective of raising standards and avoiding failures, but with a workload, a lack of resources, and a lack of management direction which meant that. . .those objectives were never going to be met (§9.53).*

Failure to remember the lessons of the past is common; I have written a book on the subject[5]. Lessons are learnt after an incident but often forgotten after a few years, when staff have changed. Before Clapham Junction the lesson that there was a need for better signalling installation and testing was never really learnt, probably because, as at Aberfan (Chapter 13), the consequences of errors had not been serious, though it was obvious that they might have been. Clapham Junction was waiting to happen.

The report criticised the clarity of British Rail rules and the language used (§11.49–11.57). Although the report does not say so the semi-legal language gives the impression that the rules are written to protect the writer rather than help the reader.

19.5   Other comments

The report advocates some strengthening of railway carriages so that they can better withstand an accident (§15.33–15.36). There has, of course, been a great improvement in their strength since the days of wooden coaches, which disintegrated in an accident and easily caught fire. The report is therefore rather cautious, suggesting further research to see what can be done.

The report makes many suggestions (in Chapter 5) for improving the response of the emergency services. These are not criticisms, the emergency services are praised, but detailed recommendations on communications, exercises, etc.

The report made 93 recommendations. Unlike the report on the fire at King's Cross, they are not prioritised. As in the King's Cross report there is no quantitative risk assessment or cost-benefit analysis and no comparison of the recommendations with other ways of using the nation's resources (see Section 18.4).

19.6   Senior managers' responsibilities

The report makes it quite clear that there was no deliberate decision to give safety a low priority. The same is true of the reports, official and otherwise, on almost all the incidents described in this book. Incompetence, forgetfulness and the other well-known weaknesses of human nature, not wickedness, were responsible (see Part 3 of the Introduction). In the Clapham Junction report, and in that on Aberfan (Chapter 13), this is made particularly clear.

The vital importance of this concept of absolute safety was acknowledged time and again in the evidence which the Court heard. . .there was total

*sincerity on the part of all who spoke of safety in this way but nevertheless
...The appearance was not the reality. The concern for safety was permitted to coexist with working practices which were...positively dangerous..
.It has to be said that a concern for safety which is sincerely held and repeat-
edly expressed but, nevertheless, is not carried through into action, is as
much protection from danger as no concern at all* (§17.2–4).

The senior management of British Rail is obviously responsible for
this state of affairs. As in many other organisations, they had not
realised (and no one had told them) that saying that safety was important and urging people to do better was not enough. They should have
identified the problems, agreed actions and followed up to see that
action was taken. This is normal good management, but is often not
followed where safety is concerned. As we have seen, the Southern
Region Signal and Telecommunications Engineer carried out the first
of these two actions, so far as testing was concerned, but not the third
(see Section 19.2). In learning from the past, improving communications, training and promotion policy, the problems were not even identified. No one seems to have asked, 'What prevents our safety record
being better than it is?'

The report recommends Total Quality Management, a review by
outside consultants and independent auditing. It also criticises the trade
unions: *The unions too must bear their share of responsibility for attitudes
of entrenched resistance to change which are out of place in a modern
world...The internal promotion system...is a fetter upon the acquisition of
skilled staff and is long overdue for consideration* (§16.77).

Perhaps successive governments also bear some responsibility for
continual interference in the way the railways are run and for failing to
give then sufficient resources. Stanley Hall, a former railway manager,
writes that financial constraints have 'forced BR to cut its staff to the bone
and its margins below prudent levels in an almost desperate attempt to
prove to the government that it deserves to be allowed to embark on a
major capital investment programme'[6].

British Rail was fined £250 000 (plus £55 000 costs) after the accident.
The judge said that a heavier fine would merely divert cash needed for
improvements in safety[7].

Criticisms of the senior management following a railway accident is not
entirely new. In his annual report for 1870 railway inspector Henry Tyler
criticised the London and North Western Railway, Britain's largest, and
came to the conclusion:

> that the supreme management of this company is principally responsible for
> these shortcomings, in so far as the means and appliances by which a
> maximum, and even a reasonable degree of safety may be secured, have not,
> as the general result of that management, been provided[8].

After many years these views are now widely but not universally
accepted in industry.

References

1 Reid, B., quoted in *The Daily Telegraph*, 15 June 1991.
2 Hidden, A., *Investigation into the Clapham Junction Railway Accident*, Her Majesty's Stationery Office, London, 1989.
3 Kletz, T. A., *An Engineer's View of Human Error*, 2nd edition, Institution of Chemical Engineers, Rugby, UK, 1991.
4 Health and Safety Executive, *Passenger Falls from Train Doors*, Her Majesty's Stationery Office, London, 1993.
5 Kletz, T. A., *Lessons from Disaster – How Organisations have No Memory and Accidents Recur*, Institution of Chemical Engineers, Rugby, UK, 1993.
6 Hall, S., *Danger on the Line*, Ian Allen, London, 1989, p. 126.
7 *The Daily Telegraph*, 15 June 1991.
8 Neal, W., *With Disastrous Consequences. .*, Hisarlik Press, London, 1992, p. 223.

Chapter 20

Herald of Free Enterprise

At the time of the Zeebrugge disaster, ferry interests were about to shoot a video in Switzerland on the alleged hazards [of carrying motor vehicle and people through railway tunnels]; it was hastily abandoned.

Newspaper report[1]

On 6 March 1987 the cross-channel roll-on/roll-off ferry *Herald of Free Enterprise* sank, with the loss of 186 passengers and crew, soon after leaving Zeebrugge, Belgium en route for Dover, England. The vessel sank because the large inner and outer bow doors, through which vehicles enter and leave, had been left open and water soon rose the few meters necessary for it to enter the ship. The water moved to one side and caused the ship to roll onto its side and settle on a sandbank.

The official report[2], on which this chapter is based, discussed the immediate causes of the accident, the deficiencies in design and the underlying weaknesses in the management. These are summarised in Figure 20.2.

20.1 The immediate causes

The immediate cause of the accident was the fact that the assistant bosun, who should have closed the doors was asleep in his bunk and did not hear an announcement on the loudspeakers that the ship was about to sail. This allowed the newspapers to report, with their usual superficiality, that the accident was due to human error, but there was, as we shall see, much more wrong.

Why did no one notice that the doors were open before the ship sailed? There were several reasons:

- The ferry company originally used ferries with visor doors which lift up like a knight's visor and can be seen from the bridge. When they changed to ferries with clam doors, which swing open on a vertical axis, and cannot be seen from the bridge, they failed to realise that the captain on the bridge now needed an indicator light or closed circuit TV camera to tell him that the doors were shut[3]. This is a.good illustration of the maxim that changes in design may need changes in codes of practice. We cannot relax because we are following all the rules, or accepted practices, as they may be out-of-date.

- On several occasions from 1983 onwards, after other ships in the fleet had sailed with their bow doors open, several captains asked for indicator lights to be installed. These requests were not merely ignored but were treated with contempt by head office managers. One of them wrote, 'Do they need an indicator to tell them whether the deck store-keeper is awake or sober? My goodness!!'; another wrote, 'Nice but don't we already pay someone!'; a third wrote, 'Assume the guy who shuts the doors tells the bridge if there is a problem' (§18.5 of the official report). The report stated, '. . .those charged with the manage-ment of the Company's Ro-Ro fleet were not qualified to deal with many nautical matters and and were unwilling to listen to their Masters, who were well qualified' (§21.1).
- The bosun, the assistant bosun's immediate superior, was the last man to leave the immediate vicinity of the bow doors before the ship sailed. He took a narrow view of his duties and did not close the doors himself or make sure that the man who should do so was on the job (§10.2).
- A general instruction stated that the officer loading the main vehicle deck should ensure that the bow doors were 'secure when leaving port'. This instruction was generally ignored (§10.4). The officer merely checked that the assistant bosun was on the job. On this occasion the officer saw a man approaching and thought he was the assistant bosun. The officers and crew worked different shift systems and the officer on duty could not recognise the assistant bosun (§10.8).
- The assistant bosun did not have to report that the doors were shut. The captain assumed that everything was in order unless he was told that something was not.

There were thus at least five ways in which the immediate cause of the sinking, leaving the bow doors open, could have been prevented. There were many people, from company director and designers downwards, who could have prevented it.

20.2   The weaknesses in design

To quote from a former ship's master, 'It does not require any technical knowledge to see that if an 8 000 ton ship turns upside down in 4 minutes when water enters right forward through a door 6 ft to 8 ft (1.8–2.4 m) above the waterline, in mild weather, there is something very radically wrong with the design'[4].

On most ships the cargo space is divided by bulkheads which prevent the free movement of any water that enters the ship. On ro-ro ferries bulkheads would make the loading of vehicles more difficult and would reduce the cargo space available so it became standard practice to omit them. As a result, movement of the ship can cause any water that gets onboard to move to one side and the ship then becomes unstable. The Inquiry recommended that the use of portable bulkheads should be studied (§50.2) and that longitudinal bulkheads, which cause less interference with loading than transverse ones, should be considered (§50.3).

Other suggestions included extra buoyancy (§50.4–50.5), bigger pumps for removing any water than gets onboard (§52) and more and better emergency exits.

Due to the design of the harbour at Zeebrugge the *Herald of Free Enterprise* had to back a short distance away from the berth before the bow doors could be closed. The Inquiry recommended that berths should be modified so that this was not necessary (§30). These and other recommendations for changes in design are summarised in Figure 20.1.

20.3   The weaknesses in management

After criticising the ship's Master and Chief Officer (and recommending their suspension) the report said, 'the underlying or cardinal faults lay higher up the company. The Board of Directors did not appreciate their responsibility for the safe management of the ships. They did not apply their minds to the question: What orders should be given for the safety of our ships?' and 'From top to bottom the body corporate was infected with the disease of sloppiness' (§14.1).

To support these remarks the report listed several complaints or suggestions from the ships' Masters, in addition to the suggestion to fit door indicator lights on the bridge, which fell on deaf ears. These included complaints that the ships went to sea carrying too many passengers; requests for bigger ballast pumps; and complaints that draught marks could not be read (§16).

Responsibilities at Board level were not clear. One director told the Inquiry that he was prepared to accept that he was responsible for the safe operation of the Company's ships; another director said that no director was solely responsible (§16). Referring to the Operations and Technical Directors, the report said '. . .the Court was singularly unimpressed by these gentlemen' (§14.1).

There was no system for auditing the standard of operation although in 1986 the Department of Transport recommended that 'every company operating ships should designate a person ashore with responsibility for monitoring the technical and safety aspects of the operation of ships.' (§14.2). In fact the chairman of the holding company (which had acquired the ferry company only a few months before the disaster) was reported in

RECOMMENDATIONS FOR SAFER FERRIES

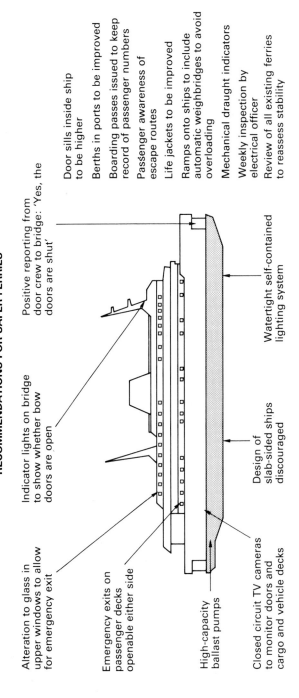

Alteration to glass in upper windows to allow for emergency exit

Emergency exits on passenger decks openable either side

High-capacity ballast pumps

Closed circuit TV cameras to monitor doors and cargo and vehicle decks

Indicator lights on bridge to show whether bow doors are open

Design of slab-sided ships discouraged

Positive reporting from door crew to bridge: 'Yes, the doors are shut'

Watertight self-contained lighting system

Door sills inside ship to be higher

Berths in ports to be improved

Boarding passes issued to keep record of passenger numbers

Passenger awareness of escape routes

Life jackets to be improved

Ramps onto ships to include automatic weighbridges to avoid overloading

Mechanical draught indicators

Weekly inspection by electrical officer

Review of all existing ferries to reassess stability

Figure 20.1 Summary of the recommendations made by the Public Inquiry for changes in the design of ferries

the press as saying that 'Shore-based management could not be blamed for duties not carried out at sea'[5]. This was true in the days of Captain Cook when ships on long voyages were not seen for months, even years, but today ships, especially those on short crossings, are as easy to audit as a fixed plant. In most of the accidents described in this book the heads of the organisations concerned have accepted responsibility after the accident (see the quotation at the beginning of Chapter 19). The exceptions are Chernobyl (Chapter 12), the *Herald of Free Enterprise* and some aircraft accidents (Chapter 21).

The Court was surprised that the officer in charge of loading did not remain on the car deck until the doors were closed. There was pressure for the ship to leave as soon as possible and the officers felt obliged to leave the deck as soon as loading was complete. But if they had stayed until the doors were closed the delay would have been only 3 minutes.

The written instructions for the officers were unclear and contradictory and this had been pointed out as early as 1982, and ignored (§11). If we do not discuss instructions with those who will have to carry them out, we end up with instructions that cannot be fulfilled. (Another example: chemical plant operators were told to add a reactant at 45°C over 30 minutes. The heater was not powerful enough for them to do so; they had either to add it more slowly or at a lower temperature [or so they believed]. They decided to do the latter and did so for many years, without the manager noticing, until ultimately it caused a fume emission.)

Unlike the Master and First Officer, the Directors were not suspended.

20.4   Learning the lessons of the past

As in most accidents[6], including those discussed in this book, there was ample opportunity to learn from previous accidents. In 1545 the *Mary Rose* sailed out of Portsmouth with her lower gunports open; she sank and most of the crew were drowned[7].

More recently, '...by the autumn of 1986 the shore staff of the Company were well aware of the possibility that one of their ships would sail with the stern or bow doors open' (see Section 20.1). Nevertheless, as already explained, they refused to fit indicator lights on the bridges, a modification so simple that after the *Herald* sank they were fitted on the remaining ships in a few days (§18.8).

20.5   Was the *Herald* typical of marine standards?

Was the accident to the *Herald of Free Enterprise* an exceptional incident or did it typify the standards prevailing in the shipping industry? The latter according to Douglas Foy[8].

In 1959, he writes, 2.8 ships in every thousand were lost; in 1979, 5.6. This worsening record is due, he says, to:

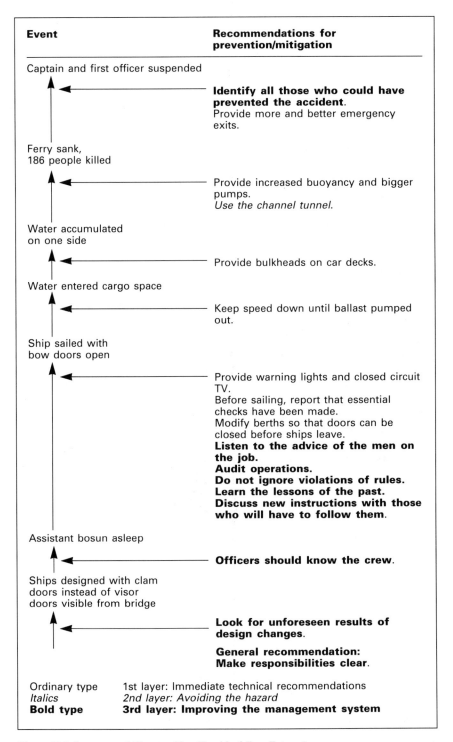

Event	Recommendations for prevention/mitigation
Captain and first officer suspended	**Identify all those who could have prevented the accident**. Provide more and better emergency exits.
Ferry sank, 186 people killed	Provide increased buoyancy and bigger pumps. *Use the channel tunnel.*
Water accumulated on one side	Provide bulkheads on car decks.
Water entered cargo space	Keep speed down until ballast pumped out.
Ship sailed with bow doors open	Provide warning lights and closed circuit TV. Before sailing, report that essential checks have been made. Modify berths so that doors can be closed before ships leave. **Listen to the advice of the men on the job.** **Audit operations.** **Do not ignore violations of rules.** **Learn the lessons of the past.** **Discuss new instructions with those who will have to follow them.**
Assistant bosun asleep	**Officers should know the crew.**
Ships designed with clam doors instead of visor doors visible from bridge	**Look for unforeseen results of design changes.** **General recommendation: Make responsibilities clear**.

Ordinary type   1st layer: Immediate technical recommendations
Italics   *2nd layer: Avoiding the hazard*
Bold type   **3rd layer: Improving the management system**

Figure 20.2 Summary of Chapter 20 – *Herald of Free Enterprise*

- Many more ships flying the flags of countries that are unwilling or unable to regulate them and leave the responsibility to classification societies that are ill-equipped to do so.
- An increase in the number of small owners who leave crew hire to agents.
- Larger ships without any corresponding improvement in manning or maintenance.
- Willingness of insurance companies to insure unseaworthy ships.
- Poor design, particularly of large ore carriers and roll-on/roll-off ferries. Only the last of these applies to the *Herald!*

Foy quotes many examples to illustrate his argument. For example, in 1979 two very large crude oil carriers collided in darkness and rain in the West Indies as the watchkeeping officers on both ships did not make proper use of their radars and did not follow the Collision Regulations. One of the ships exploded and sank as its tanks had not been inerted; one of its lifeboats was mishandled and 26 people were drowned. The master had not held a boat drill during his 10 months on the ship.

Thirty-three per cent of the world tonnage of ships flies flags of convenience but they account for 60% of accidents[9].

The Braer, which ran aground off the Shetland Isles in 1992, causing an oil spillage, was flying the flag of Liberia. The captain was Greek, the non-commissioned officers were Polish and the crew Filipino[9].

In 1993 the Polish ferry *Jan Heweliusz* sank with the loss of fifty lives while sailing from Ystad, Sweden to Swinoujscie, Poland. It was suggested that some of the cargo, which included ten railway carriages and 29 lorries had broken loose in bad weather and moved to one side but the ship had a history of accidents. It had keeled over in harbour in 1982 because its ballast tanks were wrongly adjusted and had rammed the quay the week before it capsized. Finally, it had sailed despite bad weather as the competition on the route was intense[10].

Rømer *et al.*[11] quote data on the probability of a fire, collision, etc., on a ship.

On the crossing used by the *Herald* it will soon be possible to avoid the hazard by using the channel tunnel.

References

1 Comfort, N., *Daily Telegraph*, 3 October 1989.
2 Department of Transport, *MV Herald of Free Enterprise: Report of Court No 8074*, Her Majesty's Stationery Office, London, 1987.
3 Spooner, P., *Disaster Prevention and Management*, Vol. 1, No. 2, 1992, p. 28.
4 Isherwood, J.H., letter to *The Daily Telegraph*, 14 October 1987.
5 Quoted by McIlroy, J., *The Daily Telegraph*, 10 October 1987.
6 Kletz, T.A., *Lessons from Disaster – How Organisations have No Memory and Accidents Recur*, Institution of Chemical Engineers, Rugby, UK, 1993.
7 Hewitt, P.D., letter to *The Daily Telegraph*, 13 October 1987.

8 Foy, D., *Journal of the Royal Society of Arts*, Vol. 136, No. 5377, Dec. 1987, p. 13.
9 *Free Labour World*, February 1993, p. 5.
10 *The Daily Telegraph*, 15 January 1993.
11 Rømer, H., Brockhoff, L., Haastrup, P. *et al.*, *Journal of Loss Prevention in the Process Industries*, Vol. 6, No. 4, 1993, p. 219.

Chapter 21

Some aviation accidents

Casualties in aviation are generally caused by involuntary vertical movement towards the ground.

Prospectus for shareholders in a fleet of airships[1]

This chapter applies the methods of layered accident investigation to some aircraft accidents and is based mainly on the detailed accounts in Stanley Spencer's book, *Air Disasters*[2]. The accidents are summarised in Figures 21.1–21.9 (which should be read from the bottom upwards) and the text draws attention to the salient points. As we shall see, air crash investigators are much readier than people in industry to blame the pilots (or, in one case, the controllers) instead of looking for weaknesses in designs or procedures, hence the saying, 'If the crash doesn't kill the pilot, the inquiry will'.

21.1   The R101 airship disaster (Figure 21.1)

Britain's 1920s airship programme was planned to bring comfort and speed to air travel, unite the far-flung British Empire and bring prestige to the country. It ended disastrously in 1930 with the loss of the R101 and the forty-eight people on board.

   The auguries were good. Between 1919 and 1930 all seven attempted crossings of the Atlantic by airship had been successful while 27 attempts to cross by aeroplane had resulted in 16 failures and 21 deaths. What therefore went wrong? Early flights showed that the weight of R101 was greater than expected and that the gas bags (lined with the intestines of oxen!) and their outer covering rubbed against the structure and became holed. Extensive modifications were necessary after which there was time for only one test flight (in perfect weather) before the planned prestige flight to India. The Air Minister, Lord Thomson, an airship enthusiast, had been appointed Viceroy of India, wanted to arrive there by airship and put a great deal of pressure on those concerned. 'I must insist on the

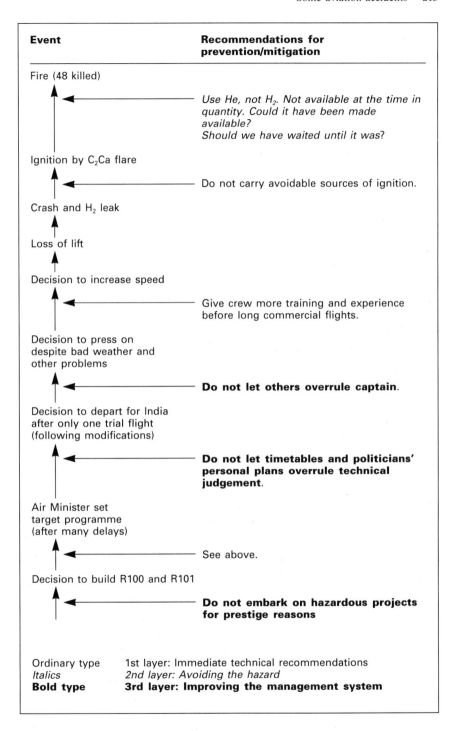

Figure 21.1 The R101 airship disaster 1930

programme for the India flight being adhered to as I have made my plans accordingly', he wrote. The Air Ministry's own inspector refused to issue a 'Permit to Fly' but was overruled.

Soon after the start of the maiden voyage one engine failed and the airship ran into strong gales and heavy winds. Even worse weather was forecast. The captain wanted to turn back but the 'officer in charge of the flight', a member of the staff of the Airship Works, persuaded him not to do so but instead to increase speed and pass through the bad weather as quickly as possible. This increased the stresses on the outer covering and gas bags, leaks occurred and the airship fell slowly to the ground. On impact a box of calcium carbide flares (which ignite on contact with water and can be dropped as flares when passing over the sea) hit the wet ground and ignited. The leaking hydrogen caught fire and soon the whole airship was alight. The forty-eight people killed included the Air Minister and senior members of the Ministry staff.

The official report was, by any standard, superficial. It merely said that the accident was due to loss of gas in bumpy conditions, compounded by a downdraught. Since every aircraft will meet bad weather sooner or later this was hardly an explanation.

At the time US airships were using helium instead of hydrogen but the supply was limited and none was available for export. National enthusiasm for the project and political pressure by the Air Minister for a success for the Labour government would have made it difficult to postpone the project until helium was available.

21.2   The Comet crashes (Figure 21.2)

The Comet, the first jet airliner, was another prestige project that ran into unexpected difficulties. The de Haviland Company realised that they could compete with US manufacturers only by innovating and they set out to develop an aircraft that would fly almost twice as fast and almost twice as high as existing machines. Despite the setback of three major crashes and several lesser ones, jet aircraft became the standard worldwide.

The story of the major crashes illustrates our reluctance to accept that anything is fundamentally wrong until the evidence becomes overwhelming. The first crash, in October 1952, in which no one was killed, and the second, in March 1953, a delivery flight, in which eleven people were killed, were blamed on pilot error. The first major crash, in India in May 1953, which killed forty-three people, was blamed on an exceptional gust of wind, so high that it would have endangered any aircraft. The wreckage was scattered over 2 km^2. Pressure tests on a fuselage showed that stress concentrations at the corners of the square windows could cause fatigue failures but the test conditions were so extreme that they could not occur in flight.

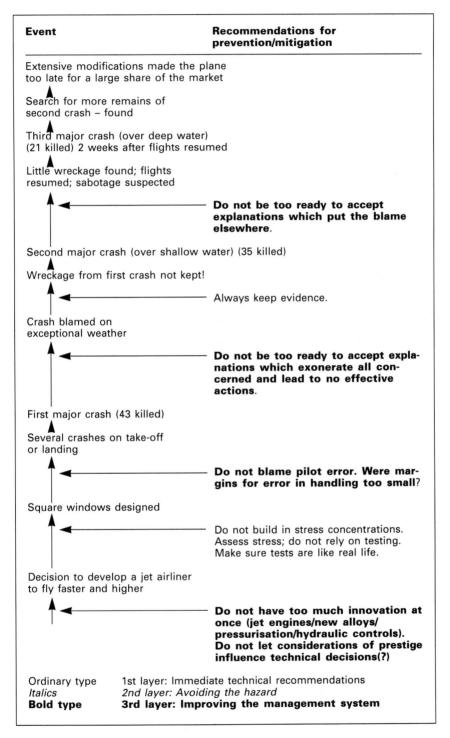

Event	Recommendations for prevention/mitigation
Extensive modifications made the plane too late for a large share of the market	
↑ Search for more remains of second crash – found	
↑ Third major crash (over deep water) (21 killed) 2 weeks after flights resumed	
↑ Little wreckage found; flights resumed; sabotage suspected	
↑ ←—————————	**Do not be too ready to accept explanations which put the blame elsewhere.**
Second major crash (over shallow water) (35 killed)	
↑ Wreckage from first crash not kept!	
↑ ←—————————	*Always keep evidence.*
Crash blamed on exceptional weather	
↑ ←—————————	**Do not be too ready to accept explanations which exonerate all concerned and lead to no effective actions.**
First major crash (43 killed)	
↑ Several crashes on take-off or landing	
↑ ←—————————	**Do not blame pilot error. Were margins for error in handling too small?**
Square windows designed	
↑ ←—————————	Do not build in stress concentrations. Assess stress; do not rely on testing. Make sure tests are like real life.
Decision to develop a jet airliner to fly faster and higher	
↑ ←—————————	**Do not have too much innovation at once (jet engines/new alloys/pressurisation/hydraulic controls). Do not let considerations of prestige influence technical decisions(?)**

Ordinary type   1st layer: Immediate technical recommendations
Italics   *2nd layer: Avoiding the hazard*
Bold type   **3rd layer: Improving the management system**

Figure 21.2 The Comet crashes 1953/4

The second major failure, in which thirty-five people were killed occurred over the sea near the Italian coast in January 1954. Again, many pieces of wreckage were spread over a wide area. Some of the wreckage was recovered but despite intensive investigations no cause could be found. Sabotage was suspected. Some minor changes were made, including fitting armour plating round the engines so that engine failure could be contained, and flights were resumed.

Two weeks later, in April 1954, a third Comet, with twenty-one people on board, disappeared over the Mediterranean. The water was too deep for recovery of the wreckage but more wreckage from the second major crash was recovered. This showed that fatigue cracks near the window corners had led to failure of the fuselage.

Why was this not spotted during the earlier tests on a complete fuselage? The fuselage had been treated as a pressure vessel and, as a safety measure, tested at a pressure above design, before the fatigue tests (using fluctuating pressures) started. This initial test relieved some of the stress and the subsequent fatigue tests were not like real life[3].

In addition, de Haviland had not calculated the stress but relied on testing. The inquiry said that they could not be criticised for doing so, given the state of knowledge at the time.

It was 1958 before a new Comet, with oval windows, was ready for service. By this time the Boeing 707 and Douglas DC-8, both with longer range, were available and although the new Comet continued to fly for 25 years its rivals had the lion's share of the market.

21.3   The Munich crash (Figure 21.3)

In 1958 a British European Airways Ambassador aircraft, taking off from Munich airport, failed to clear the end of the runway and crashed into a house less than 30 m beyond the end. Of the forty-four people on board, twenty-three were killed, including eight players and four other employees of Manchester United football team.

The weather was bad. There were several centimetres of slush on the runway and some of the other planes taking off at about the same time had been de-iced. The captain of the British plane decided that this was not necessary and the German inquiry blamed the accident on his error of judgement.

At the time the effect of slush on acceleration was not fully understood or the runway might have been cleared. As evidence accumulated, from experiments and from reports by other pilots, that slush might have been responsible, there were calls for the inquiry to be re-opened but the German authorities refused. Finally, ten years after the accident a British Commission cleared the captain, who had survived. One of their witnesses was the first man on the scene after the crash; he had examined

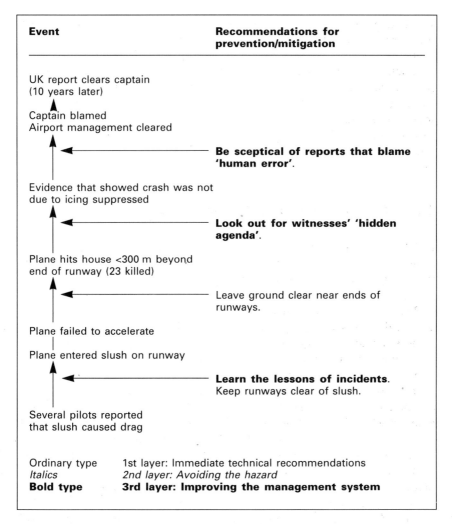

Figure 21.3 The Munich crash 1958

the wings and found no ice on them. He had not been called to the German inquiry, only someone who detected a patch of ice six hours later.

One of the lessons we can learn from the Munich accident, therefore, is the need to look, in all investigations, for hidden agenda. We all tend to blame other people and it is easy for manufacturers or, as in this case, airport managements, to blame the crew. It seems that in this case evidence was suppressed and key witnesses were not called though the airport management could have argued that, with the knowledge available at the time, they were justified in ignoring the slush. Self-protection of the sort that occurred at Munich is not only unjust; as it does not establish the cause of an accident, we cannot take action to prevent it happening again.

A somewhat similar accident occurred at Athens airport in 1979. The runway was coated with oil and rubber and the first rain for a long period had made it slippery. An aircraft overran and fell down a 4 m drop. For seven years pilots had been asking the airport to fill in the drop, without success. Nevertheless, the pilots were prosecuted for manslaughter and sentenced to imprisonment, but released on appeal.

21.4   The Trident Crash at Heathrow (Figure 21.4)

In 1972 a British European Airways Trident crashed soon after leaving London's Heathrow Airport. All the 118 people on board were killed. The triggering event was the sort of human error that is sooner or later inevitable. The underlying cause was a weakness in design.

Rear-engined aircraft such as the Trident are inherently liable to stall. The Trident was well supplied with automatic stall warning and recovery devices including a 'stick shaker' which shook the control column at an early stage and a 'stick pusher' which moved the control column forward, at a later stage. There were also various warning lights and pilot training covered the recognition and handling of approaching stalls. Protective equipment to prevent stalling, however, is inherently less satisfactory than a design which is less liable to stall. In addition, teething troubles had left pilots with the impression that the systems were unreliable.

The Trident wings were fitted with flaps and also with droops (on their front edges). If the droops were retracted too soon after take-off, the plane was liable to stall. Flaps were normally retracted first and the droops a few minutes later. On some occasions flaps were retracted sooner than usual and then the pilot, out of habit, would select the only lever available and inadvertently retract the droops at the time when he would normally retract the flaps. This had happened on at least two occasions but the pilots had recovered the situation. On the day of the crash they did not.

At the time many pilots were on strike and an inexperienced pilot was rostered as second officer. The captain had had a row with his colleagues before joining the plane and suffered a heart attack during take-off. With a combination of illness and inexperience the crew were unable to prevent the stall.

It is not helpful to say that the accident was due to human error, made worse by illness and inexperience. The error was a result of poor design. Designers should realise that operators cannot always perform perfectly and design accordingly. We all have days when, for one reason or another, we do not perform as well as we normally do.

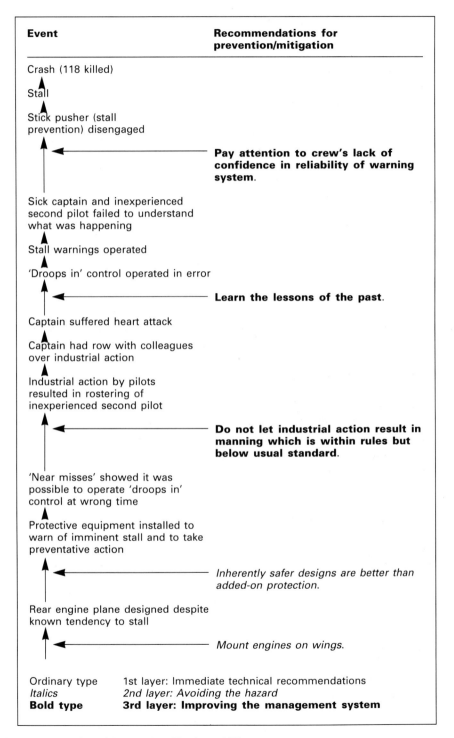

Figure 21.4 The Trident crash at Heathrow 1972

21.5   The Paris DC-10 crash (Figure 21.5)

The crash of a Turkish Airlines DC-10 near Paris in 1974 was the culmi-
nation of a long story of hushing up and patching up. By 1972 over a
hundred reports of faults in the door closing mechanism of the DC-10 had
been sent to the manufacturers, McDonnell-Douglas, who advised the
airlines to fit more powerful actuators. They did not tell the US Federal
Aviation Authority (FAA); they were not required by law to do so but
manufacturers usually reported this sort of thing. Then, in 1972, a cargo
door flew open during flight and a hole appeared in the floor of the
passenger cabin (as the pressure in the cabin was higher than outside and
the floor was not strong enough to withstand the difference).

The National Transportation Safety Board (NTSB), an advisory body,
asked for further modifications, including a more foolproof design of
closing mechanism. The FAA, weakened by politicians, made recommendations only and they did not go as far as the NTSB wanted. They did,
however, ask for a peephole or viewing port by the door so that the man
closing it could see if the catch was home. A diagram by the door told
him what to look for. As in the last incident, protective devices did not
remove the hazard; they merely allowed someone to check that the door
was secure.

Because of an oversight by McDonnell-Douglas the modifications to a
Turkish Airlines DC-10 were not completed correctly though the view
port was fitted. This aircraft stopped in Paris on the way to London.
British European Airways were involved in a strike and transferred many
of their Paris–London passengers onto the Turkish flight; it left with 346
people on board. The cargo door was closed by a man who could not read
English and did not check the view port. When the plane crashed everyone on board was killed.

21.6   The Zagreb mid-air collision (Figure 21.6)

This accident is an appalling example of the way some managements try
to hide their own responsibility by blaming those at the bottom of the
organisation who cannot pass the blame on to someone else.

In 1976 the control equipment at Zagreb airport was poor and the
controllers were overworked. There had been thirty-two near-miss collisions in the past five years. A relief controller was late arriving and the
senior controller left his seat for eight minutes in order to look for him,
leaving an assistant in charge of the section. As the result of a misunder-
standing during the brief handover, compounded by a missing arrow on
the screen which should have shown that an aircraft was climbing, two
aircraft collided in mid-air with the loss of 176 lives.

All the controllers on duty were put on trial for criminal misconduct
but all except the assistant were acquitted. He was sentenced to seven

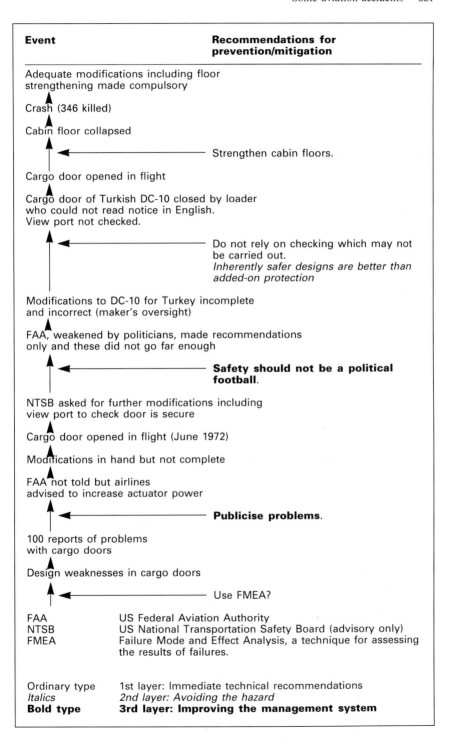

Figure 21.5 The Paris DC-10 crash 1974

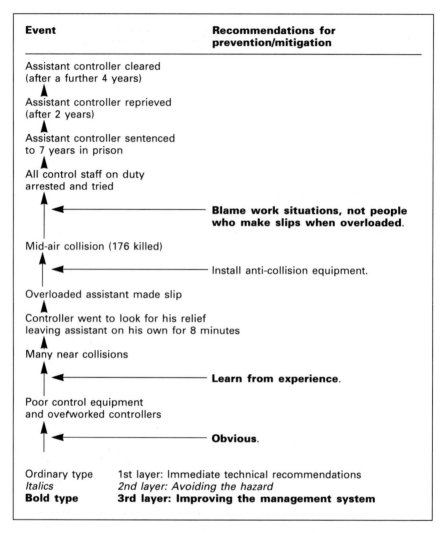

| Event | Recommendations for prevention/mitigation |

Assistant controller cleared
(after a further 4 years)
▲
Assistant controller reprieved
(after 2 years)
▲
Assistant controller sentenced
to 7 years in prison
▲
All control staff on duty
arrested and tried
▲ ◄———————————— **Blame work situations, not people who make slips when overloaded.**

Mid-air collision (176 killed)
▲ ◄———————————— Install anti-collision equipment.

Overloaded assistant made slip
▲
Controller went to look for his relief
leaving assistant on his own for 8 minutes
▲
Many near collisions
▲ ◄———————————— **Learn from experience.**

Poor control equipment
and overworked controllers
▲ ◄———————————— **Obvious.**

Ordinary type	1st layer: Immediate technical recommendations
Italics	*2nd layer: Avoiding the hazard*
Bold type	**3rd layer: Improving the management system**

Figure 21.6 The Zagreb mid-air collision 1976

years imprisonment, halved on appeal. He was released after two years and in 1982 a re-convened investigation agreed that the assistant had been overloaded.

21.7   The Tenerife ground collision (Figure 21.7)

In March 1977 two 747 Jumbo jets collided on the ground at Los Rodeos Airport, Tenerife in the Canary Islands. Of the passengers and crew on the two planes 583 were killed and only seventy survived.

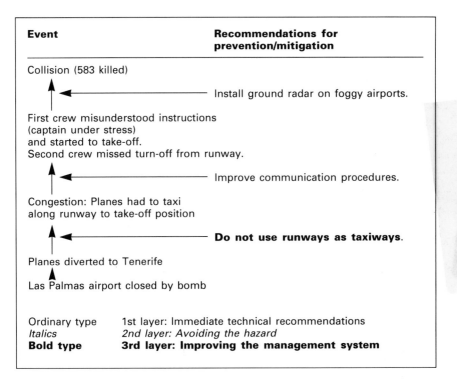

Event	Recommendations for prevention/mitigation
Collision (583 killed)	
↑ ← ───────────	Install ground radar on foggy airports.
First crew misunderstood instructions (captain under stress) and started to take-off. Second crew missed turn-off from runway.	
↑ ← ───────────	Improve communication procedures.
Congestion: Planes had to taxi along runway to take-off position	
↑ ← ───────────	**Do not use runways as taxiways**.
Planes diverted to Tenerife ↑ Las Palmas airport closed by bomb	

Ordinary type	1st layer: Immediate technical recommendations
Italics	*2nd layer: Avoiding the hazard*
Bold type	**3rd layer: Improving the management system**

Figure 21.7 The Tenerife ground collision 1977

The main airport in the Islands, Las Palmas, had been closed following a bomb explosion and planes were diverted to Los Rodeos. The airport was congested and planes had to taxi along the runway to reach their parking or take-off positions instead of using the normal taxiway parallel to the runway. Fog made visibility poor.

Soon after Las Palmas was re-opened a KLM plane was waiting at one end of the Los Rodeos runway for clearance to take off for Las Palmas while a Pan Am plane was taxiing along the runway from the other end, intending to take a turn-off onto the taxiway. In the fog the Pan Am pilot missed the turn-off and continued along the runway. At the same time the KLM pilot misunderstood his instructions and started to take off. A message from the Pan Am pilot to the controller, 'OK, we'll report when we're clear', overheard by the KLM crew, did not register. The two planes collided about halfway along the runway.

Why did the experienced Dutch pilot misunderstand? According to Stewart[2]

> ...the flying environment, though for the most part routine, can place great strain on an individual...The Dutch crew had been on duty for almost 9½ hours and still had to face the problems of the transit in Las Palmas and the return to Amsterdam...The pressure was on to leave Los Rodeos as soon

as possible and the weather did not help. . .Close concentration was required
on the take-off as the clouds were again reducing visibility. At such moments
the thought process of the brain can reach saturation point and can become
overloaded. The 'filtering effect' takes over and all but urgent messages, or
only important details of the task in hand, are screened from the mind.
Radio communications, which were being conducted by the first officer,
were obviously placed in a low priority in the minds of both pilots once the
take-off had been commenced. The controller's use of papa alpha instead
of Clipper [in addressing the Pan Am crew] – the only occasion on that day
on which that identification was used – reduced the chances of registering
that transmission.

Using the runway as a taxiway is obviously bad practice, particularly
when crews cannot see whether or not it is clear, and was used only
because the airport was overcrowded. Once again we see people trying to
control a hazard by procedures which are subject to occasional but
inevitable misunderstandings and mishearings, rather than avoiding the
hazard (by sticking to separate runways and taxiways) or installing better
control equipment (ground radar).

21.8   The Chicago DC-10 crash (Figure 21.8)

In May 1979 an engine fell off an American Airlines DC-10 while it was
taking off from Chicago Airport. All 271 people on board and two people
on the ground were killed. The immediate cause was fatigue cracking on
an engine pylon flange. Checks showed similar damage on other DC-10s
and also discrepancies in various clearances and failed or missing fasten-
ers.

The story started in 1974 when pylon assembly was moved from one
factory to another, skilled men were lost and quality fell. In 1975 the
manufacturers, McDonnell-Douglas, asked the airlines to carry out certain
maintenance work on the pylons. Instead of dismantling the engine *in situ*,
as recommended by the manufacturers, American Airlines, amongst
others, removed the pylon assembly as a whole using a fork lift truck. This
was cheaper and quicker though the manufacturers said they would not
encourage the procedure. They did not specifically forbid it.

In 1978 a pylon flange failed on a Continental Airlines DC-10 during
maintenance. A report on this incident was circulated to other airlines but
it was camouflaged by the trivia of other incidents and did not make it clear
that the failure was related to the method used for removing the pylon.

Following the 1979 crash the US Federal Aviation Authority grounded
all DC-10s for five weeks while further examinations were carried out.

When an engine fell off in flight the power supply to various instruments and warning systems was lost, the pilots did not know that a stall
was imminent and were unable to prevent it. Perhaps more redundancy
in the power and warning systems is needed.

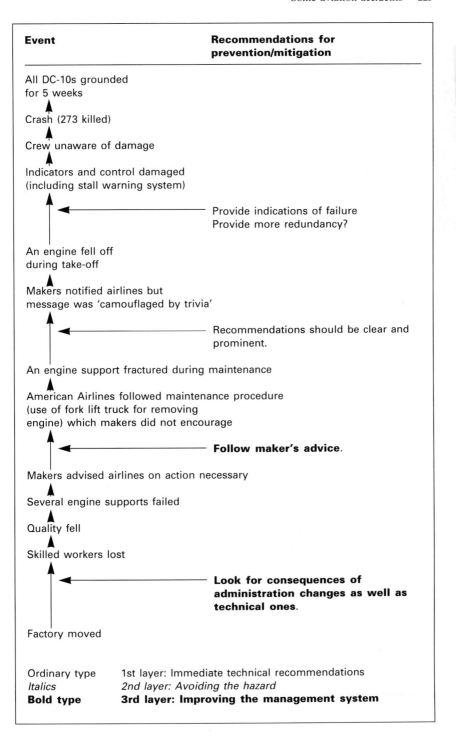

Figure 21.8 The Chicago DC-10 crash

21.9   The Mount Erebus crash (Figure 21.9)

Witnesses are sometimes reluctant to tell all they know and, as we have seen in Section 21.3, evidence is occasionally suppressed, but deliberate lying is rare. One of the few cases in which this undoubtedly occurred has been described by Peter Mahon, the judge in charge of the inquiry[4].

An Air New Zealand plane set off on a sightseeing tour of Antarctica. Unknown to the captain and crew the destination waypoint to which the plane was flying on automatic control had been altered in the plane's computer. As a result the plane, which was flying at a low altitude so that the passengers could see the view, flew up the wrong valley and crashed into the rock wall at the end. All the 257 people on board were killed. The valley looked like the one the plane should have flown up. The pilot and crew had never flown on this route before but an experienced Antarctic explorer was on board, to describe the scenery to the passengers, and he failed to spot the error. The New Zealand regulatory authority had such confidence in the airline that they rubber-stamped everything proposed and did not query the use of inexperienced crews.

The crew did not see the approaching end of the valley, due to the well-known phenomenon of white-out which makes near mountains disappear leaving what appears to be a distant horizon. By the time the low altitude warning sounded it was too late to avoid the ground ahead. There was no forward-facing ground proximity system, something that might have been considered for low level flying. However, the whole concept of a low level sightseeing flights was flawed, as it involved quite unnecessary risks.

The airline management tried to blame the captain, saying that he should not have flown so low, though every other flight had done so and the company's own newspaper made it quite clear that the flights took place at a low height. Mahon describes in fascinating detail how he gradually came to realise that he was being told 'an orchestrated litany of lies'.

21.10   Challenger

In January 1986 the US space shuttle *Challenger* was destroyed by fire soon after take-off and the crew of seven were killed. The prestige of NASA, the US National Aeronautics and Space Administration, suffered a serious blow, from which it took a long time to recover, and the US space programme was delayed for several years.

The story is summarised in Figure 21.10, though some of the events occurred simultaneously rather than in the order indicated. It started about 1970 when President Nixon decided to go ahead with the construction of a re-usable space shuttle. Unlike President Kennedy he was not a space enthusiast and decided to back a project that, he hoped, would bring him credit during his term of office, rather than a long-term project such as an expedition to Mars. About the same time many of the senior staff

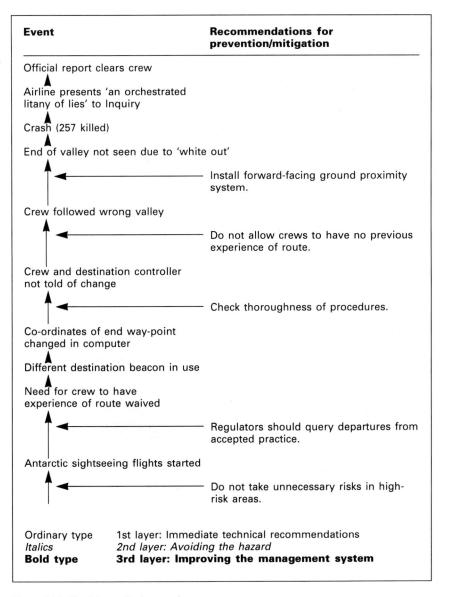

Figure 21.9 The Mount Erebus crash

at NASA, who were political appointees, were changed, so that the organisation lost experience and continuity[5].

NASA accepted the cheapest quotation for the design of the shuttle even though the casing would be made in parts. One of the four quotations was for a one-piece design, an inherently safer one as its integrity does not depend on a additional pieces of equipment, seals, which are

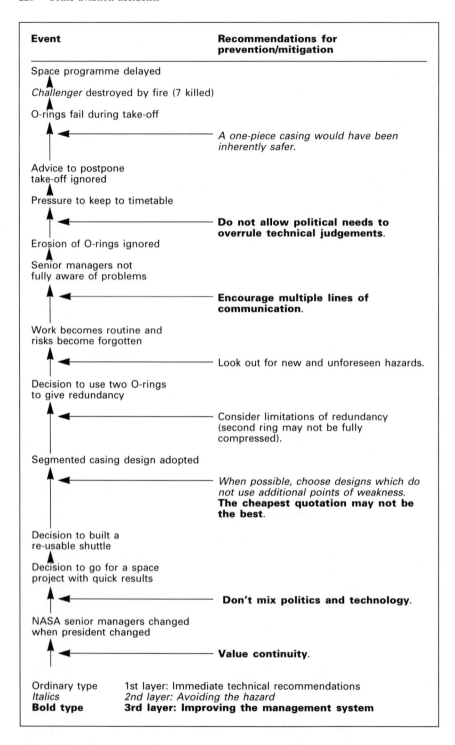

Event	Recommendations for prevention/mitigation
Space programme delayed	
↑	
Challenger destroyed by fire (7 killed)	
↑	
O-rings fail during take-off	
↑ ◄────────	*A one-piece casing would have been inherently safer.*
Advice to postpone take-off ignored	
↑	
Pressure to keep to timetable	
↑ ◄────────	**Do not allow political needs to overrule technical judgements.**
Erosion of O-rings ignored	
↑	
Senior managers not fully aware of problems	
↑ ◄────────	**Encourage multiple lines of communication.**
Work becomes routine and risks become forgotten	
↑ ◄────────	Look out for new and unforeseen hazards.
Decision to use two O-rings to give redundancy	
↑ ◄────────	Consider limitations of redundancy (second ring may not be fully compressed).
Segmented casing design adopted	
↑ ◄────────	*When possible, choose designs which do not use additional points of weakness.* **The cheapest quotation may not be the best.**
Decision to built a re-usable shuttle	
↑	
Decision to go for a space project with quick results	
↑ ◄────────	**Don't mix politics and technology.**
NASA senior managers changed when president changed	
↑ ◄────────	**Value continuity.**

Ordinary type   1st layer: Immediate technical recommendations
Italics   *2nd layer: Avoiding the hazard*
Bold type   **3rd layer: Improving the management system**

Figure 21.10 *Challenger*

inevitably weaker in nature than the the parts they join. The failure of one of these O-ring seals led to the loss of *Challenger*.

Sections 21.4 and 21.5 described accidents which occurred because designers relied on procedures instead of removing a hazard. In the space shuttle they relied on equipment which was inherently weak.

Realising that the O-rings were weak features, the designers decided to duplicate them. However, this did not give them the redundancy they thought it would as one ring in a pair may be gripped less tightly than the other one[6].

The first flights showed that the O-rings became eroded but this caused little concern among the senior management. Development of a new design, a 27-month programme, was started but meanwhile the old design continued in use, for several reasons[6]:

- Many years of success had dulled the feeling that space travel was a risky enterprise; it had become almost routine: the *Titanic* syndrome: 'We don't need lifeboats as the ship can't sink'.
- There was pressure to cut costs and keep to programme.
- Senior managers were not fully aware of the problems, a statement they often make after an accident. 'If I had known what was happening I would have stopped it.' This is no excuse. Managers should not wait to be told of problems but should find them out for themselves.

NASA was a structured organisation in which people reported problems to their bosses but felt they could do nothing if their bosses ignored the problems; they did not want to risk their reputations by going directly to more senior managers. The fault lay with the culture, which is, of course, the responsibility of senior managers. It is fatal to rely solely on one's immediate subordinates and formal lines of communication for information; every manager should tour his parish and chat to people at all levels.

Finally, on the day before the fatal launch several engineers urged that it should be postponed as the temperature was below freezing and the O-rings would be brittle and more likely to fail. They were overruled, for the same three reasons. Several writers have suggested that the pressure to keep to programme came from President Reagan, as he was due to make a State of the Union Message later that day and would have liked to report a successful start to the flight. However, it is equally or more likely that the senior staff at NASA *thought* they were being pressured.

After the accident many of the engineers who had opposed the flight were punished while the managers survived. The company who had designed the shuttle earned extra payment for redesigning it.

References

1   Quoted by Green, M., *The 'Peterborough' Book*, David and Charles, Newton Abbot, UK, 1980, p. 3.
2   Stewart, S., *Air Disasters*, Ian Allan, London, 1986.

3 Taylor, D.A.W., *Design for Safe Operation – Experimental Rig to Production*, Symposium Proceedings No 5, Institute of University Safety Officers, 1985, p. 4.
4 Mahon, O., *Verdict on Erebus*, Collins, Auckland, New Zealand, 1984. For shorter accounts see reference 2, p. 172 and Shadbolt, M., *Readers Digest*, November 1984, p. 164. For the full story see *Report of the Royal Commission to inquire into the Crash on Mount, Erebus, Antarctica of a DC10 Aircraft operated by Air New Zealand Limited*, Government Printer, Wellington, New Zealand, 1981.
5 Trento, J.,J., *Prescription for Disaster*, Crown, New York, 1987.
6 Bell, T.E. and Esch, E., *IEEE Spectrum*, February 1987, p. 36.

Chapter 22

Conclusions

It is a proverbial remark that we in England never begin making a reform or adopting an improvement till some disaster has demonstrated its absolute necessity.

The Illustrated Times, 19 January 1867[1]

This chapter summarises those recommendations that have appeared in many of the preceding chapters. Note that many important recommendations are not included as they appear in only one or two of the preceding chapters. For example, errors in the preparation of equipment for maintenance, a major cause of accidents, appear only in Chapters 2, 5 and 17. The accidents described were chosen primarily to illustrate the need for analysis in depth. Many accidents due to the most important immediate technical causes are described in reference 2.

22.1 Effective prevention lies far from the top event

In many of the accidents discussed too much reliance was placed on measures which were designed to prevent the upper events in the diagrams (or the upper events in fault trees). If the measures failed, for any reason, there was then little or no opportunity for further defence.

Chapter 4 describes the most extreme example. Everyone was casual about leaks of ethylene because they had, they thought, eliminated all causes of ignition and the leaks could not, therefore, ignite. When a source of ignition appeared, an explosion was inevitable. Similarly, at King's Cross (Chapter 18) nobody worried about escalator fires as they were sure they could be easily extinguished. Those concerned were in the position of a military commander who neglects all the outer lines of defence because he considers the inner stronghold impregnable.

Bhopal (Chapter 10) is another example. There was extensive provision for dealing with high pressures in vessels containing methyl isocyanate (MIC): a cooling system to lower the temperature and thus the pressure, relief valves, a scrubbing system and a flare system. When the cooling, scrubbing and flare systems were not in full working order, a discharge to atmosphere was inevitable once a trigger turned up. A more effective loss prevention programme would have avoided an intermediate stock of MIC, kept the public away from the plant and tackled the underlying reasons that lay behind the failure to maintain the safety equipment – such as the lack of training and education.

It may be useful to summarise the main lines of defence that can be used to prevent major leaks of hazardous materials, starting with those that lie furthest from the top event. Some apply only to flammable materials.

(1) Avoid large inventories of hazardous materials by intensification, substitution or attenuation (see Section 22.4).
(2) Inspect thoroughly during and after construction (see Chapter 16).
(3) Install gas detectors so that leaks are detected promptly (see Chapter 4). This does not remove the need for regular tours of inspection by operators. Even on plants which are fitted with gas detectors about half the leaks that occur are detected by men.
(4) Warn people when a leak occurs. Those who are not required to deal with a leak should leave the area, by a safe route (see Chapters 4 and 5).
(5) Isolate the leak by means of remotely-operated emergency isolation valves (see Chapter 4).
(6) Disperse the leak by open construction supplemented, if necessary, by steam or water curtains (see Chapter 4). (It may be better to contain leaks of toxic gas unless they will disperse before they reach people such as members of the public or workers on other plants who are not trained to deal with them.)
(7) Remove known sources of ignition. Though this may seem a strong line of defence it is actually one of the weakest. As we have seen in Chapter 4, so little energy is needed to ignite a leak of flammable vapour and air that a source of ignition is liable to turn up even though we think we have removed all such sources.
(8) Protect against the effects of the leak, as follows:
Fire: Insulation and water spray.
Explosion: Strengthened buildings,
                 distance (i.e. prevent development nearby).
Toxicity: Distance (i.e. avoid concentrations of people nearby).
(9) Provide fire-fighting and other emergency facilities.

This list, of course, takes no account of the software measures such as hazard and operability studies, audits, etc., some of which are discussed below.

22.2   The control of plant modifications

Many of the accidents (see Chapters 1, 7, 8, 14 and item 6 of Chapter 15) occurred because modifications of plant and process had unforeseen and undesirable side-effects and the actions that should be taken to prevent similar accidents in the future were summarised in Section 7.1. (See also references 1 and 2 of Chapter 7.) People dealing with complex systems tend to think in straight lines. They think about the effects of their actions on on the path to the immediate goal but remain unaware of the side-effects[3].

22.3   Testing and inspection of protective equipment

Many of the accidents would not have occurred if protective equipment, of various sorts, had been kept in working order (see Chapters 1, 2, 3, 6, 7, 10, and 12). To prevent similar accidents in the future we need a two-pronged approach:

(1) An education programme to convince people, at all levels, that safety equipment can and should be kept in working order and that it is not an optional extra, something that can be neglected or put to one side under pressure of work (see Chapter 6). A one-off programme after an accident is not sufficient. An ongoing programme is necessary. An occasional lecture or piece of paper is not sufficient. It is better to involve people in regular discussions of accidents that have occurred, why they occurred and the action necessary to prevent them happening again (see part 4 of the Introduction), and to circulate regular reminders, in well-written and attractive publications, of the accidents that have happened and the precautions necessary. Compare your company's safety literature with that prepared to attract your customers! The safety adviser also has something to sell.

Training in loss prevention is particularly important during a person's formative years as a student. Such training is standard practice in the UK but not in most other countries[4].

(2) An audit or inspection programme to make sure that equipment is being kept in working order. Much equipment should be tested at regular intervals, instruments monthly, relief valves every year or two, some equipment more often, for example the vents in Chapter 7. Managers should make spot checks from time to time and there should be occasional audits by outsiders[5].

If protective equipment has to be disarmed, this should be signalled clearly so that everyone is aware of the fact, and is constantly reminded (see Chapter 3). In some cases it should be difficult or impossible to disarm the protective features (see Chapter 12).

22.4   User-friendly designs

'Friendly' is used in the computer sense to describe plants which will tolerate departures from ideal operation or maintenance without an accident occurring. Thus Bhopal would not have occurred if there had not been a large, unnecessary intermediate stock of MIC. The accident described in Chapter 6 would not have occurred if the recovered raw material had not been stored but fed straight back into the plant. 'What you don't have, can't leak (or explode).' We should keep stocks of hazardous materials to the minimum (intensification), use safer materials instead (substitution) or use the hazardous materials under the least hazardous conditions (attenuation)[6].

The plant described in Chapter 4 had numerous crossovers between parallel streams, providing many opportunities for leaks (and for errors). The closed compressor house magnified the effects of any leaks.

In the unit described in Chapter 2, the protective system was neglected and thus introduced a greater hazard than the one it was designed to remove, but the need for the system could have been avoided by moving the unit a few metres.

The nuclear reactors at Three Mile Island (Chapter 11) and Chernobyl (Chapter 12) were less friendly than gas cooled reactors as they were dependent on added-on cooling systems which were liable to fail and gave the operators less time in which to respond. Chernobyl was particularly unfriendly as it had a positive power coefficient (i.e. as it got hotter, the rate of heat production increased).

The casing of the *Challenger* space shuttle was not made in one piece but in several segments with joints between them. The O-rings in the joints were a weak feature and they failed (see Section 21.10). Rear-engined aircraft such as the Trident are inherently more liable to stall than those with the engines mounted on the wings (see Section 21.4). The cargo door catch on the DC-10, even after modification, was unfriendly as someone had to look through a view hole to make sure that it was correctly closed (see Section 21.5).

22.5   Carry out hazard and operability studies

Many of the accidents described show the need for critical examination of the design by hazard and operability studies (hazops) or similar techniques (see Chapters 8–10). The technique is now well known[7,8] and widely used and there is therefore no need to describe it here. Less widely recognised is the need for similar (but shorter) studies in the earlier stages of design if we are to avoid some of the unfriendly features just described. Two such early studies are needed, one at the conceptual or business analysis stage when we are deciding what process to use and where the plant is to be located, and one at the flowsheet stage[6]. A conceptual study at Bhopal, for example, might have queried the choice of product (other insecticides are available), the process to be used and the need for intermediate stocks. It might also have asked if development of pest-resistant plants or the introduction of natural predators would be better than insecticides. (I am not suggesting that they are – both these suggestions have disadvantages – only that the question should be asked.)

A flowsheet study would have allowed people to query the need for intermediate stocks again, the capacity of the scrubbing, flare and refrigeration systems and the need for sparage. The final hazop, on the line diagrams, would have allowed the team to explore ways in which contaminants might enter the MIC system.

Many companies will argue that they would have discussed all these subjects. This is true, but what is lacking in many companies is a structured, systematic, formalized technique in which every possible alternative (in the earlier studies) or deviation (in the final hazop) is considered in turn. Because modern designs are so complicated we cannot foresee the effects of alternatives, or deviations, unless we go through the design, bit by bit, slowly and systematically.

Samuel Coleridge described history as 'a lantern on the stern', illuminating the hazards the ship has passed through rather than those that lie ahead. It is better to illuminate the hazards after we have passed through them than not illuminate them at all, as we may pass the same way again, but it is better still to illuminate the hazards that lie ahead. This book is a lantern on the stern. Hazop is a lantern on the bow.

To use a different metaphor, like the chameleon we need to keep one eye on the past and one on the future[9].

22.6   Better management

Some necessary, but sometimes neglected, management features have already been discussed: audits, regular testing of protective equipment and hazard and operability studies.

In the incidents described in Chapters 1, 6, 7, 8, 10, 11, 12 and 19 operator training, or rather the lack of it, was a significant factor. Operators do not just need training in their duties but also need an understanding of the process and the hazards so that they can handle problems not foreseen when the instructions were written. Managerial ignorance was illustrated in Chapters 4, 5, 8 and 14.

A common feature has been failure to learn from the experience of the past. Sometimes the knowledge was forgotten, sometimes it was not passed on to those who needed to know. Sometimes the whole company or factory failed to learn (see Chapters 4, 5, and 13), sometimes individual managers (see Chapters 2 and 7). The people involved in an accident do not forget but after a while they leave and others take their place. Organisations have no memory. Only people have memories and they leave[10].

There is much that can be done to learn from the experience of those who have been involved in accidents and to keep alive the memory of the past. For example, we can:

- Discuss accidents from time to time, as described in Part 1 of the Introduction.
- Describe old accidents as well as recent ones in safety bulletins and discuss them at safety meetings.

- Include, in standards and codes of practice, notes on the the accidents that led to the recommendations.
- Keep a 'black book' or 'memory book' in each control room, a folder of reports on accidents that have occurred. Do not include falls and bruises, but only accidents of technical interest, and include accidents from other plants. The black book should be compulsory reading for newcomers, at all levels, and old hands should dip into it from time to time to refresh their memories.
- Make more use of information storage and retrieval systems so that if we become interested in, say, explosions in compressor houses, it is easier to find reports on incidents that have occurred and the recommendations made.

These recommendations are discussed more fully in reference 10.

As with the need to keep protective equipment in working order, we need more than an occasional one-off action; we need an ongoing programme.

Many of the accidents had a big impact on the public (see Chapters 8–12) and emphasise the need for hazardous industries to explain their hazards, and the precautions taken, rather better than they have done in the past if public disquiet is not to lead to restrictions on their activities.

Underlying these various specific recommendations, the culture or climate of a company affects the way the staff behave, the action they take to prevent accidents and the enthusiasm with which they take it, as Section 3.5 made clear. Different companies, even different plants within the same company, can have quite different cultures. The culture depends to some extent on the training the staff have received, both in the company and during their formative years as students, as discussed in Chapter 10, and on the examples set by senior managers. Changing the culture is difficult, and takes time, but it can be done[10].

One of the finest US loss prevention engineers, the late Bill Doyle, used to say that for every complex problem there is at least one simple, plausible, wrong solution[11]. This book confirms his opinion. It has shown that there is no 'quick fix' to the problem of preventing accidents. Each accident is different and complex; there is no single cure for the lot; each requires action at various levels, action to prevent the events that occurred immediately beforehand, action to remove the hazard and action to improve the management system. The first and third are usually possible on every plant but removing the hazard is often difficult on an existing plant. How far should we go? This is discussed in the next section.

22.7   How far should we go in bringing old plants up to modern standards?

There is no simple answer to the question but a number of examples will show how the problem can be approached[12]. If we consider extreme examples first, the question is easy to answer. Some features cannot possibly be added to old plants while others are as easy to install on an old plant as on a new one.

Standards of plant layout have improved in recent years and it is now normal to leave spaces between the sections of a large plant, or between small independent units, to restrict the spread of fire and to provide access for fire-fighting vehicles (and maintenance equipment). Little or nothing can be done to improve the layout of existing plants and if the layout is unsatisfactory all we can do, short of closing down the plant, is to compensate for the poor layout by higher standards of fire protection. Similarly, we now try to locate plants away from residential areas but housing has grown up around some old plants. Short of removing the houses or plants there is nothing that we can do.

Strengthened control rooms have been built on many new plants but it is difficult to replace an existing control room (though it has been done on one or two plants). However, we can remove large picture windows and heavy light fittings, and use special glass in any remaining windows (or glass protected by plastic film).

In contrast, gas detectors are as easy to install on an old plant as on a new one. Improved fire insulation is almost as easy to install on an old plant. Remotely-operated emergency isolation valves are rather more difficult to install, as lines have to be cut and welded, but have been successfully fitted on many old plants.

The ventilation of many old compressor houses has been improved by knocking down some or all of the walls. They are not usually load bearing. The most difficult part of the operation may be relocating equipment which is fixed to the walls.

Improvements to the drainage are more difficult. On many old plants the ground is sloped so that spillages run towards sources of ignition (see Chapter 3) or collect underneath the plant. It is often difficult or impossible to regrade the slope of the ground, but this has been done on some plants. Some old plants (see Chapter 5) have surface drains. Installing underground drains is very expensive and disruptive and is rarely attempted.

A liquid-phase hydrocarbon processing plant, actually the scene of the accident described in Chapter 5, was constructed using pipework with compressed asbestos fibre (caf) gaskets. After a number of flange leaks had occurred a change to spiral-wound gaskets was suggested but the size of the task – there were thousands of joints – appeared daunting. Further consideration led to a practicable solution. It was decided to replace all the gaskets in lines carrying liquid but not those in vapour lines. Three thousand gaskets were replaced over two years, during the normal 6-monthly shutdowns, joints exposed to liquids above their boiling points being changed first. This example shows how an improvement to an existing plant which at first appears too large to contemplate becomes practicable when it is pruned a little and spread over several years.

A similar example is provided by item 10 of Chapter 15. After a screwed joint on an old plant had leaked many, but not all, screwed joints were replaced. In general screwed joints attached to main process lines were replaced but those beyond the first isolation valve, which could be isolated if they leaked, were left alone.

Some older plants contain mild steel pipes which can become too cold, and therefore brittle, under abnormal operating conditions and this has caused some serious leaks and explosions[13]. Complete replacement of the mild steel lines is obviously impracticable. One company carried out a detailed study of its older pipework to determine its 'fitness for purpose'. Some lines were replaced; some were radiographed, to confirm that the welds were free from defects, and then stress-relieved, thus making them able to withstand lower temperatures. In other cases additional low-temperature alarms or trips were installed or the importance of watching the temperature was emphasised in operator training.

After two pipe failures had occurred another plant carried out a comprehensive review of their pipework to identify the changes needed and to put them in order of priority[14].

Replacement of relief valves which are now known to be undersized is straightforward but enlargement of the flare header into which the valves discharge is more difficult. Trip systems have therefore been used to avoid or reduce the need for larger relief valves or to avoid a corresponding increase in the size of the flare header[15].

Although we cannot always bring the hardware on our old plants up to the latest standards ('Inconsistency is the price of progress') we can make sure that the software (that is, the methods of operation, training, instructions, testing and inspection, and auditing) are fully up to standard. This is at least half the problem. The earlier chapters have shown that if we want to prevent accidents, changes to the software are just as necessary as changes to the hardware. Safety by design is a desirable objective, and should always be our aim, but is not always possible, or practicable.

The following accident shows what can happen when good procedures do not compensate for poor design.

In 1973 a cage collapsed at Markham Colliery, Derbyshire, UK, killing eighteen men and seriously injuring eleven. The design was poor. The newspapers said that sixteen separate safety systems had to fail before the cage would collapse but all these systems operated the same brake. This brake was applied by powerful springs, held off by compressed air. The springs acted through a steel rod, 2.7 m long by 50 mm diameter which operated a lever (Figure 22.1). The bearing at the bottom of the rod could not be lubricated; it became stiff, the rod became bent and then broke.

A similar brake had failed 12 years earlier, though without serious consequences. Replacement was considered too expensive and, in any case, would have taken some time. The Coal Board's Divisional Chief Engineer issued an instruction that all similar rods should be examined. He did not say how they should be examined or how often. At Markham

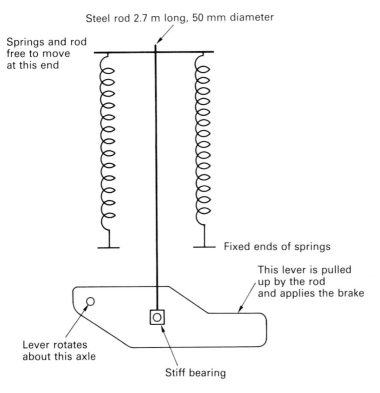

Steel rod 2.7 m long, 50 mm diameter

Springs and rod
free to move
at this end

Fixed ends of springs

This lever is pulled
up by the rod
and applies the brake

Lever rotates
about this axle

Stiff bearing

Figure 22.1 The brake system on the cage at Markham Colliery. The rod was 'single-line' and when it failed the cage fell out of control

the rod was not removed but merely examined in position and it was not scheduled for regular examination in the future. The brake was tested every 3 months but the loads used were estimated and were sometimes less than the normal working load[16].

To sum up this section:

(1) If a change is as easy, or almost as easy, to make on an old plant as on a new one, then it should be made, unless the plant is due to be shut down in a few years or there is another way of achieving the same objective.
(2) We should not accept a higher standard of risk (to operators or the public) on an old plant than we would accept on a new one, but the methods we use to obtain that standard may be different. We may have to rely more on improvements to the software. If we do decide to do so, this must not be mere lip-service. We must be prepared to demonstrate that we have really improved the software.
(3) Newer plants often get the best managers and operators. Yet old plants, with their greater dependence on high standards of operation

and management, may be in greater need of them. New plants have the glamour, and people like to be transferred to them, but we must take care that old plants are not starved of resources.

(4) When new standards are designed to prevent damage to plant or loss of production rather than injury to people, it is legitimate to leave old plants unchanged. Whether or not we do so should depend on the balance between probabilities and costs.

(5) We need to know how far our plants fall short of modern standards. Hazard and operability studies and relief and blowdown reviews can tell us.

22.8 Human error

Many accidents, as we have seen, are blamed on human error, usually an error by someone at the bottom of the pile who cannot blame someone below him. Managers and designers, it seems, are either not human or do not make errors. In the process industries this attitude is much less common than in was; many companies now realise that almost every accident is due to a management failing and they will not accept human error as the cause of an accident. Only in Chapter 12 (Chernobyl), Chapter 20 (*Herald of Free Enterprise*) and Chapter 21 (aircraft accidents) did we find senior people blaming those below them. The pilots of an aircraft, though highly skilled people, are in a similar position to process operators: they have no one below them to blame.

The term human error is not a useful one as it does not lead to effective action. If we say that an accident was due to human error, we tend then to say that someone should take more care. This is not very helpful as no one is deliberately careless. If we look more closely at so-called human error accidents we can divide them into different groups which call for different actions.

- *Mistakes*: someone did not know what to do (or, worse, thought they knew but were wrong). We should improve the training or instructions or, better still, simplify the job.
- *Violations*: someone knew what to do but decided not to do it. We should explain why the job should be done the correct way as we do not live in a society in which people will follow the rules just because they are told to do so. In addition, we should check from time to time to see that the correct methods are being used. Incorrect methods can usually be followed for months or years before an accident occurs (as at Clapham Junction, Section 19.1). (How many times have you exceeded a speed limit without being stopped?) Better still, can the job be simplified? If the safe way of doing it is difficult, people will find an unsafe way.
- *Mismatches* between the job and the ability of the person asked to do it: someone knew what to do and intended to do it but it was beyond

his or her physical or mental ability, perhaps beyond anyone's ability. In some cases we may have to change the person but it is usually better to change the work situation, that is, the plant or equipment design or the method of working.

- *Slips and lapses of attention*: someone knew what to do, intended to do it and was able to do it but failed to do it or did it incorrectly. These are like the slips and lapses of everyday life but more serious in their consequences. Human nature being what it is, they are inevitable from time to time and there is little that we can do to prevent them except, when we can, to reduce stress and distraction. It is usually more effective to change the work situation so that there are fewer opportunities for errors or so that the errors have less serious consequences (see Section 22.4).

The various sorts of human error can be illustrated by an accident which occurred because someone failed to open (or close) a valve. It could have been a mistake: he did not know he was expected to close the valve. It may have been a violation: he knew he was expected to close the valve but decided it was unnecessary or, in an extreme case, could not be bothered. It may have been a mismatch: someone intended to close the valve but found that it was too stiff. Finally, and most likely, it could have been a slip or lapse of attention: someone intended to close the valve but being busy or under stress, it slipped his or her mind.

The various sorts of human error are discussed and illustrated more fully in reference 17.

If then accidents are due to managerial failings, failure to provide adequate training, instruction, supervision, equipment or methods of working, should managers be punished after an accident? The public, encouraged by the press, seem to believe that there must be someone to blame for every accident (see quotation by A. Cruikshank at the beginning of Chapter 12). However, managers at all levels, like the rest of us, need training, instructions and supervision and like the rest of us, do not always get them. Often directors and senior managers do not provide them because they themselves do not realise that they could do more to prevent accidents by providing better leadership for their staff. It is not sufficient to exhort people to work safely and to comment on the lost-time accident rate. They should identify the real problems, agree solutions and follow up to see that the solutions are working. This is, of course, normal good management but is often not followed so far as safety is concerned. Safety training for those at the top of a company is more important than training for those at the bottom as if those at the top have the right approach, the rest will follow automatically.

An official report pointed out that most management schools do not include safety in their curricula and that most books on management do not mention it[18].

22.9   A final note

I have, I hope, established the main thesis of this book, that most accidents
have no simple cause but arise as the result of many interacting circumstances. As a result there are many ways of breaking the chain of events
that culminates in an accident. To what extent is this true elsewhere? Can
we apply layered investigation in other fields?

Many writers have shown that the effects of natural disasters are not
entirely due to natural causes but are often made worse by human actions.
Thus removing trees which absorb water can make land prone to flooding; removing vegetation which stores water can make land prone to
drought[19]. To say that the floods or droughts are due to Acts of God,
beyond our control, is at best simplistic, at worst wrong.

Similarly, famines are not simply due to harvest failure (nor is harvest
failure always due to climatic irregularities or crop disease). Most famines
occur when there is no absolute food shortage, but rather a breakdown of
distribution facilities or a shortage of money with which to buy the food
that is available. The great Bengal famine of 1943 was triggered by a 5%
fall in food production. Conversely, harvest failure has little effect if food
can be supplied from elsewhere. To prevent famine we have to look at
public and private provision for relief, transport, food storage and much
more. Finally, is famine the major cause of starvation or is ongoing malnu-
trition a greater problem?[20]

Similarly diseases have many causes, as Hippocrates pointed out many
centuries ago[21]. It is simplistic to ask if a particular illness is due to hered-
itary or the environment, to infection or something else. Susceptibility to
a disease may be inherited but it may not develop unless other factors are
present as well.

In an entirely different field, why was there an enormous growth in
cathedral and church building in the 13th century? In France 600 cathe-
drals and major churches were built between 1170 and 1270.

The immediate technical causes may have been the invention of the
ribbed vault and rising living standards which made available resources
above those needed for subsistence. But these merely made the construction of large churches possible. Why did those in authority want to build
them? It was an age of faith; perhaps they believed that anything was
possible with God's help, and wished to raise monuments to His Glory.
But the 13th century was also an age of questioning and heresy, new ideas
and better communications, an age in which people wanted to do what
had not been done before, and these factors may have been more important[22].

A final word of caution: suppose an accident is due to corrosion. Such
a statement, I have tried to show, is superficial. Before we can prevent
further corrosion we need to know much more. Was corrosion foreseen
during design and if not, why not? Were the right materials of construction used? Was there any monitoring for corrosion, and, if so, was any

notice was taken of the results? Were operating conditions the same as those foreseen during design?

The statement about corrosion, though superficial, is probably correct. If we have doubts, it is easy to get another opinion. As we uncover the underlying causes it often becomes more difficult to be sure that we have uncovered the truth. We may be shown an instruction that certain corrosion monitoring should be carried out but was this instruction known to the people on the job? Did anyone check to see that it was followed? Was it practicable to follow it? Were the necessary resources provided? We must not allow ourselves to be fobbed off by statements such as 'the accident was due to organisational weaknesses'. This may be OK as an overall conclusion but organisations have no minds of their own. Such statements should be backed up by recommendations that named people should carry out specific actions. Unless this is done, nothing will happen, except a repeat of the accident.

References

1 Quoted by Neal, W., *With Disastrous Consequences. . .*, Hisarlik Press, London, 1992, back cover.

2 Kletz, T. A., *What Went Wrong?* 2nd edition, Gulf Publishing, Houston, Texas, 1988.

3 Reason, J., *Bulletin of the British Psychological Society*, Vol. 40, April 1987, p. 201.

4 Kletz, T. A., *Plant/Operations Progress*, Vol. 7, No. 2, April 1988, p. 5.

5 Kletz, T. A., *Lessons from Disaster – How Organisations have No Memory and Accidents Recur*, Institution of Chemical Engineers, Rugby, UK, 1993, Chapter 7.

6 Kletz, T. A., *Plant Design for Safety – A User-Friendly Approach*, Hemisphere, New York, 1991.

7 Kletz, T. A., *Hazop and Hazan*, 3rd edition, Institution of Chemical Engineers, Rugby, UK, 1992.

8 Knowlton, R. E., *A Manual of Hazard and Operability Studies*, Chemetics International, Vancouver, Canada, 1992.

9 Malagasy proverb, quoted by A. Jolly, *National Geographic Magazine*, Vol. 171, No. 2, Feb. 1987, p. 183.

10 Kletz, T. A., *Lessons from Disaster – How Organisations have No Memory and Accidents Recur*, Institution of Chemical Engineers, Rugby, UK, 1993.

11 W. H. Doyle was probably paraphrasing H. L. Mencken, 'There is usually an answer to any problem: simple, clear and wrong' (quoted by Rosten, L., *Leo Rosten's Book of Laughter*, Elm Tree Books, London, 1986, p. 363).

12 Kletz, T. A., *Loss Prevention*, Vol. 14, 1981, p. 165.

13 van Eijnatten, A. L. M., *Chemical Engineering Progress*, Vol. 73, No. 9, Sept 1977, p. 69.

14 Matusz B. T. and Sadler, D. L., 'A comprehensive program for preventing cyclohexane oxidation process piping failures', *Proceedings of the 27th Annual Loss Prevention Symposium*, American Institute of Chemical Engineers, New York, 1993, Paper 12f.

15 Kletz, T. A., *Improving Chemical Industry Practices – A New Look at Old Myths of the Chemical Industry*, Hemisphere, New York, 1990.

16 *Accident at Markham Colliery, Derbyshire*, Her Majesty's Stationery Office, London, 1974.

17 Kletz, T. A., *An Engineer's View of Human Error*, 2nd edition, Institution of Chemical Engineers, Rugby, 1991.

18 Advisory Committee on Safety of Nuclear Installations Human Factors Study Group, *Third Report: Organising for Safety*, Her Majesty's Stationery Office, London, 1993, p. 47.

19 Wijkman, A. and Timberlake, L., *Natural Disasters – Acts of God or Acts of Man*, International Institute for Research and Development, London, 1984, pp. 6, 29 and 30.
20 Garnsey, P. in *Understanding Catastrophe*, edited by Bourriau, J., Cambridge University Press, Cambridge, UK, 1992, p. 145.
21 Wingate, P., *The Penguin Medical Encyclopaedia*, Penguin Books, London, 1972, p. 208.
22 Tuchman, B., *Practising History*, Ballantine Books, New York, 1982, p. 232.

Appendix

Some questions to ask during accident investigations

This questionnaire is intended to help us think of some of the less obvious ways of preventing an accident happening again. It will be more effective if the questions are answered by a team rather than a person working alone. The team leader should be reluctant to take 'Nothing' or 'We can't' as answer. He should reply, 'If you had to, how would you?'

WHAT is the purpose of the operation involved in the accident?
  WHY do we do this?
  WHAT could we do instead?
  HOW else could we do it?
  WHO else could do it?
  WHEN else could we do it?
  WHERE else could we do it?

WHAT equipment failed?
  HOW can we prevent failure or make it less likely?
  HOW can we detect failure or approaching failure?
  HOW can we control failure (ie, minimise consequences)?
  WHAT does this equipment do?
  WHAT other equipment could we use instead?
  WHAT could we do instead?

WHAT material leaked (exploded, decomposed, etc.)?
  HOW can we prevent a leak (explosion, decomposition, etc.)?
  HOW can we detect a leak or approaching leak (etc.)?
  WHAT does this material do? Do we need so much of it?
  WHAT material could we use instead?
  WHAT safer form could we use the original material in?
  WHAT could we do instead?

WHICH people could have performed better? (Consider people who might supervise, train, inspect, check, or design better than they did as well as people who might construct, operate and maintain better than they did.)

WHAT could they have done better?

HOW can we help them to perform better? (Consider training, instructions, inspections, audits etc as well as changes to design.)

Afterthought

Man has three ways of learning: Firstly, by meditation; this is the noblest. Secondly, by imitation; this is the easiest. Thirdly, by experience. This is the bitterest.

Confucius

Applied to hazards, *meditation* includes systematic techniques such as hazard and operability studies and design calculations.

Imitation includes learning from the experience of others, as described in design standards, codes of practice, books and accident reports.

Experience is waiting until we have had an accident ourselves[1].

---

[1] Based on the Foreword to *Safety Management: The Hong Kong Experience*, by Lee Hung-Kwong, Lorrainelo Concept Design, Hong Kong, 1991.

Index

DATE DUE			